Multi-scale Spectral Analysis in Hydrology

Multi-scale Spectral Analysis in Hydrology

From Theory to Practice

Adarsh S and M Janga Reddy

CRC Press
Taylor & Francis Group
Boca Raton London New York

CRC Press is an imprint of the
Taylor & Francis Group, an informa business

First edition published 2021
by CRC Press
6000 Broken Sound Parkway NW, Suite 300
Boca Raton, FL 33487-2742

and by CRC Press
2 Park Square, Milton Park, Abingdon, Oxon OX14 4RN

Library of Congress Cataloging-in-Publication Data
[Insert LoC Data here when available]

ISBN: 9780367622015 (hbk)
ISBN: 978-0-367-62206-0 (pbk)
ISBN: 9781003108351 (ebk)

Visit the [companion website/eResources]: [insert CW/eResources URL]

Dedication

To Our Parents

Contents

Preface

Let the time series speak for itself ...Time-series characterization is quite essential for capturing the behavior of the series and for the efficient forecasts. Proper understanding of the nonlinear and nonstationary features is an essential prerequisite for the effective simulation and prediction. The spectral analysis methods are suitable tools for time-series characterization and the classical methods like Fourier spectral analysis displays shortcomings in performance, if the time-series signal is complex nonlinear (NL) and nonstationary (NS) in characteristics. The Fourier methods of spectral analysis converts the signal from time to frequency domain and may suffer from compromise in quality, on dealing with the simultaneous transformation of complex NL-NS signals like hydrological signals. In this context, the advanced spectral analysis methods like Wavelet transforms (WT) and Hilbert Huang Transform (HHT) are found to be appropriate for the time-frequency (TF) characterization of complex signals. Both of these transforms facilitate multiscale feature extraction through effective decomposition of the candidate time series to signals of specific periodicity. The discrete version of wavelet transform found specific applications like trend analysis, hybrid predictive modeling while continuous variants are more popular in periodicity estimation and teleconnection studies in hydrology. HHT was proposed as a complementary tool to WT, which was reported to be successful in dealing with the inherent complexities of selection of appropriate mother wavelet and decomposition level, during the utilization of WT for T-F characterization of complex signals. The so called criticism on empirical mode decomposition (EMD) phas of HHT, such as lack of strong mathematical background can be considered as an advantage, as it allows the *time series to speak for itself* by resulting in decomposition of specific number of modes including the trend, even without fixing the decomposition levels *a priori*. Within two decades of its introduction, many theoretical advancements have reported, including robust noise-assisted variants of EMD, advanced Hilbert transform algorithms, HHT based running correlation analysis method like Time Dependent Intrinsic Correlation (TDIC) and Multivariate Empirical Mode Decomposition (MEMD) facilitating the simultaneous decomposition of multiple time-series signals. However, the enormous potential of these HHT-based methods in hydrology are not well debated in literature. The theoretical development of advanced spectral analysis methods like WT and HHT facilitate the use of such techniques for multiscale spectral characterization in hydrology and hence helps in overcoming the general critique on lack of real field practical applications of spectral analysis methods in hydrology. This book bridges this lacunae between theory and practical applications of spectral analysis methods in hydrology, by demonstrating number of case studies on trend analysis, hydroclimatic teleconnections, developing hydrologic frequency tool, fractal characterization, simulation of hydrological variables of different spatio-temporal scales, etc.

MATLAB® is a registered trademark of The MathWorks, Inc. For product information, please contact:

The MathWorks, Inc.
3 Apple Hill Drive
Natick, MA 01760-2098 USA
Tel: 508 647 7000
Fax: 508-647-7001
E-mail: info@mathworks.com
Web: www.mathworks.com

Acknowledgments

We would like to express our deep sense of gratitude to our family members for their invariable support, love, and affection. We express gratitude to the anonymous reviewers who made positive remarks on this book. We express our special and sincere thanks to Prof. Francois G. Schmitt, Former Director, Laboratory of Oceanology and Geosciences (LOG), University of Lille, Wimereux, France, for the scientific discussions on the Hilbert-Huang Transform held at LOG in October 2014. We acknowledge the service of faculties and staff of Department of Civil Engineering, IIT Bombay, who helped in the preparation of this book.

Above all, we would like to express a deep sense of gratitude to GOD Almighty for making our dream a reality.

About the Authors

Dr. Adarsh S, currently working as an associate professor with the Department of Civil Engineering, TKM College of Engineering Kollam, Kerala India, is a leading researcher in statistical hydrology. He completed his Ph.D. at the prestigious Indian Institute of Technology Bombay (IIT Bombay) India in 2018 and received the *Excellence in Research Award*. His specific research interests include stochastic hydrology, application of artificial intelligence techniques in hydrology, hydroclimatology, water resources systems, risk/ uncertainty analysis in hydrology. He published 55 papers in international journals of high reputation, 4 book chapters, 1 book, and over 50 papers in national and international conferences. He supervised 7 Master's theses, 43 undergraduate projects, successfully completed 3 minor research projects and supervised 2 doctoral scholars. Dr. Adarsh is currently an editorial board member of *Springer Nature Applied Sciences* and an active reviewer of more than 35 prestigious journals published by ASCE, Elsevier, Wiley, Spinger, IEEE, Taylor & Francis etc. He received the outstanding reviewer recognition by ASCE in 2019 and the Sivapalan Young Scientists Travel Award (SYSTA) in 2020–2021. He is a member of the International Commission on Statistical Hydrology (ICSH-IAHS), Institution of Engineers (India), Indian Society for Technical Education (ISTE), European Water Resources Association (EWRA), Indian Meteorological Society (IMS), Indian Water Works Association (IWWA) Indian Association of Hydrologists (IAH), etc.

Dr. M Janga Reddy is currently working as Associate Professor with the Department of Civil Engineering, Indian Institute of Technology, Bombay. He obtained his Ph.D (Engg) at the Indian Institute of Science, Bangalore in 2007. His research interests include hydrology and water resources, hydrological modeling, reservoir systems, irrigation, hydropower, flood control planning; water supply systems-pipe networks, canals; simulation and optimization modeling; risk assessment of hydrological extremes floods; and droughts. He has published over 60 papers in peer-reviewed journals and 70 papers in proceedings of national and international conferences. Dr. Reddy guided several students for their dissertations at IIT Bombay (4 Ph.D. and 14 M.Tech. theses). He is a member of the national and international professional bodies–IAHS (International Association for Hydraulics Research, UK), IE (Institution of Engineers, India), IWWA (Indian Water Works Association, India), and ISH (Indian Society for Hydraulics, India). He is an active reviewer of more than 35 international journals (published by ASCE, IEEE, Wiley, Springer, IAHS, etc.).

1 Introduction

1.1 BACKGROUND

Modeling of different components of hydrological cycle such as precipitation, evaporation, streamflow, etc. are important in water resources planning and management. The real field data of most of the hydrological variables are represented as a time series, which are often nonlinear and nonstationary in characteristics. More specifically, the changes in statistical moments or covariance over the time domain refers to the nonstationarity of time series, while nonlinearity refers to data series possessing features such as asymmetric cycles, bimodality, nonlinear relationship between lagged variables, time irreversibility, and sensitivity to initial conditions (Fan and Yao 2003). Stationary assumption is the fundamental rationale in hydrological modeling, but such an assumption may lead to wrong estimates when applied to prediction problems in hydrology under the climate change scenario. This forces researchers to revisit the traditional practices of modeling of hydrological processes accounting for nonstationarity. Spectral analysis techniques can be used to get better insight into the characteristics of time series, by the determination of the frequency content of a time series, inference of the physical mechanisms responsible for this frequency content, evaluation of the performance of simulation models (Fleming et al. 2002). Such information may eventually lead to improvements in the prediction of hydrological variables in a nonstationary environment.

1.2 SCOPE OF MULTISCALE SPECTRAL ANALYSIS IN HYDROLOGY

The Fourier spectral analysis (FSA) has been traditionally used for the spectral characterization of a hydrological time series. But the FSA is suitable only for analyzing stationary time series, as there may be significant loss of information during the transformation to the frequency domain. However, most of the signals in practical hydrology contain numerous nonstationary characteristics such as drift, trends, abrupt changes, etc., and the FSA is not suitable to detect such features. One solution to this issue is to decompose the time series into time-frequency space. The Short Time Fourier Transform (STFT) introduced by Dennis Gabor (Gabor 1946) is helpful to map a signal into time-frequency space. In this method, the Fourier transform is used to analyze only a small section of the signal at a time (popularly known as windowing of the signal); but it was noted that the precision with which the information obtained is highly dependent on the size of the window chosen (Misiti et al. 2008). Overcoming the shortcomings of the Wavelet Transforms (WT) was evolved in 1980s, which include discrete and continuous variants. Both classes of wavelets finds numerous applications in the domain of time series analysis and their applications drawn considerable attention by the researchers in hydrology for the past two decades (Sang 2013; Nourani et al. 2014). Wavelet transforms can be used to extract short-term and long-term fluctuations by decomposing the time series into different subcomponents. From past literature, it is noticed that the multiscale decomposition of time series using wavelets are useful in understanding underlying character of the hydrological processes, which may thus help to forecast the hydrological variables accurately (Nourani et al. 2014).However, the choice between a large- or small-scale wavelet function and the selection of appropriate wavelet functions and decomposition levels are reported to be two major challenges in the use of wavelet transforms for the spectral analysis of a time series (Sang et al. 2016). In addition, the results obtained by wavelet analysis will be accurate only if the nonstationary series is linear in nature. By addressing these shortcomings, Huang et al. (1998) proposed a novel data adaptive decomposition method namely Empirical Mode Decomposition (EMD) and combined it with the traditional Hilbert Transform (HT) to develop a new spectral analysis method, namely, Hilbert-Huang Transform (HHT). By using HHT, the complexity in selection of appropriate mathematical functional form and decomposition level can be addressed. Over the two decades after its introduction, HHT is established as a potential tool for spectral analysis of a complex time series including a hydrologic series. The estimation of dominant periodicities of hydrological time series and climatic oscillations, extraction of nonlinear trend of hydrological series, hydroclimatic teleconnections in multiple time scales, use of single and multivariate EMD for decomposition of hydroclimatic signals into different time scales and its application in simulation or prediction of hydrological variables by hybrid methods involving data-driven methods, etc. are some possible domains where HHT has been applied. Both the wavelets and HHT are capable of providing the information in multiple time scales, which is an added advantage in prediction and feature extraction problems in hydrology. The information on multiscaling behaviors of different complex hydrological processes could help in improved hydrological predictions. The prediction of hydrological variables could be improved by finding the possible association of hydrologic processes with climate indices having specific periodicity. Such challenging issues can be handled in a better way by following an efficient multiscale decomposition process coupled with an appropriate spectral analysis technique. But most of the past studies investigated such associations based on computation of periodicities alone. However, in order to investigate the association between two time series having multiscale characteristics in a better way, a technique which enables a running correlation analysis in multiple time scales is more appropriate. One such technique that works based on the HHT is the Time Dependent Intrinsic Correlation (TDIC) method, which can be explored for hydroclimatic teleconnection studies. It is also well known that multiple variables may influence the hydrological processes, but multiscale decomposition methods such as EMD or its variants are univariate in nature and decomposition of multiple variables of concern using such methods may not give the same number of modes. As a result, at a specific time scale, the frequency content pertaining to the modes of different variables may be different, which may lead to erroneous interpretations in teleconnection studies. To rectify such problems the teleconnection analysis can be performed effectively using Multivariate Empirical Mode Decomposition (MEMD) method, here the common scales present in multiple variables of concern can be identified in a single step operation.

The hydroclimatic variables often possess multiscaling character. Therefore, to model the hydrological processes, a decomposition-based technique may be a better alternative. Eventhough many hybrid decomposition models were proposed for simulation and prediction of hydrologic variables, only few of them considered multiple variables in the modeling process. Also, many such studies considered appropriate lags at different time scales in the modeling exercise. Moreover, many of the decomposition-based hybrid models have used the decomposed components directly as inputs, by which the significant information from specific process scales are not accounted in modeling. In this context, the potential of MEMD can be utilized in modeling, as it facilitates to account both multiple inputs and associated features in multiple time scales. In addition, MEMD-based decomposition can be used as a useful mean to determine the representative scaling exponent of rainfall intensity series of different durations, which in turn may help in develop the hourly rainfall intensity duration frequency (IDF) relationships from longer duration rainfalls (such as monthly/daily) by using the scaling theory. In short, the usefulness of advanced spectral analysis methods such as WT or HHT and its algorithmic variations needs to be investigated in the context of characterization, teleconnection, and prediction of hydrological variables. This clearly bridges the gap between theoretical principles and practice. In this perspective, this book gives a comprehensive presentation of such practical frameworks along with the demonstration through a number of case study applications in the Indian context.

1.3   ORGANIZATION OF THE BOOK CONTENT

Chapter 2 first provides the brief theoretical description of conventional spectral analysis methods followed by detailed description of advanced spectral analysis methods such as wavelet transform and HHT. The descriptions of single and multivariate EMD, recent algorithmic developments like Arbitrary Order Hilbert Spectral Analysis (AOHSA) and Time Dependent Intrinsic Correlation (TDIC) are also presented in the chapter. Three novel frameworks for hydrological applications, the MEMD-TDIC approach for multiscale teleconnection, the MEMD-scaling theory approach for developing rainfall intensity-duration-frequency (IDF) curves, and the MEMD-based hybrid modeling for simulation and prediction of hydrological variables, are presented in the chapter. Chapter 3 is devoted for wavelet transform applications for hydrologic characterization. The extraction of trend using Discrete Wavelet Transform (DWT) and application of Continuous Wavelet Transform (CWT) for teleconnection are two major applications considered in this chapter. Chapter 4 considers HHT applications on rainfall time series. This chapter covers the time-frequency characterization, teleconnections, trend analysis and development of IDF curves of rainfall. Analysis of multiscale teleconnections of streamflow with sediment load and climate variables, fractal characterization using HHT, etc. are described in Chapter 5. Chapter 6 is exclusively devoted to the simulation of different hydrological time series such as rainfall, streamflow, and suspended sediment using MEMD-based hybrid models.

REFERENCES

Fan, J., Yao, Q. 2003. Nonlinear Time Series: Nonparametric and Parametric Methods. New York: Springer-Verlag.

Fleming, S.W., Lavenue A.M., Aly, A.H., Adams, A. 2002. Practical applications of spectral analysis to hydrologic time series. *Hydrological Processes* **16**: 565–574.

Gabor, D. 1946. Theory of communication. *Journal of IEEE* **93**: 429–457.

Huang, N.E., Shen, Z., Long, S.R., Wu, M.C., Shih, H.H., Zheng, Q., Yen, N.C., Tung, C.C., Liu, H.H. 1998. The empirical mode decomposition and the Hilbert spectrum for non-linear and non-stationary time series analysis. *Proceedings of Royal Society London, Series A* **454**: 903–995.

Misiti, M., Misiti, Y., Oppenheim, G., Poggi, J.M. 2008. *MATLAB User's Guide: Wavelet Toolbox 4*. Natick, MA: Math Works Inc.

Nourani, V., Baghanam, A.H., Adamowski, J., Kisi, O. 2014. Applications of hybrid wavelet artificial intelligence models in hydrology: A review. *Journal of Hydrology* **514**(6): 358–377.

Sang, Y.F. 2013. A review on the applications of wavelet transform in hydrology time series analysis. *Atmospheric Research* **122**(2013): 8–15.

Sang, Y., Singh, V.P., Sun, F., Chen, Y., Liu, Y., Yang, M. 2016. Wavelet-based hydrological time series forecasting. *Journal of Hydrologic Engineering* 10.1061/(ASCE)HE.1943–5584.0001347, 06016001.

2 The Theory of Advanced Spectral Analysis Methods

2.1 BACKGROUND

The spectral analysis tools can be used to characterize the time series signals and understand the processes involved. The classical Fourier spectral analysis is perhaps the most popular among these tools. But its efficacy in performance is limited to linear and stationary time series while the practical hydrologic time series rarely possess the properties of linearity or stationarity. The introduction of Short Time Fourier Transform (STFT) put forwarded the concept of time and frequency localization, but the constant and *a priori* fixation of window size was a problem for the modeler to work with the technique. Also it was rather found to be a difficult task to maintain the quality of localization in one of the domain without compromising the quality of localization in the other domain during the implementation of STFT. Overcoming such limitations Wavelet transforms (WT) evolved as a potential alternative and the Continuous Wavelet Transform (CWT) and Discrete Wavelet Transform (DWT) have received attention in various applications in processing hydroclimatic time series data. But the difficulties in choosing the appropriate wavelet type and level along with its inferior capabilities in handling non-stationarity lead researchers for a data adaptive decomposition method. Hilbert Huang Transform (HHT) is one such multiscale spectral analysis method suitable for time-frequency characterization of nonlinear and nonstationary time series. HHT first finds appropriate inputs for Hilbert transform by performing a multistage decomposition process called Empirical Mode Decomposition (EMD) to evolve orthogonal subseries called Intrinsic Mode Functions (IMFs) from a given time series data. Then the IMFs are subjected to Hilbert transform to give instantaneous amplitudes and instantaneous frequencies. This chapter presents a brief information on Fourier transform and wavelet transform followed by the detailed theoretical description of EMD, its noise-assisted variants, its multivariate extension, Hilbert Spectral Analysis (HSA), the procedure of HHT based Time Dependent Intrinsic Correlation (TDIC) analysis, etc. The chapter also gives the details of proposed methods such as TDIC based approach for hydroclimatic teleconnection studies, MEMD-Stepwise Linear Regression (SLR) hybrid model for time series prediction and MEMD-based procedure for developing rainfall Intensity-Duration-Frequency relationships.

2.2   CONVENTIONAL SPECTRAL ANALYSIS METHODS

2.2.1   FOURIER TRANSFORM

Fourier transform can be viewed as a transformation in function space from the time domain to the frequency domain which contains trigonometric functions like *sines* and *cosines* as basis functions that are localized in frequency only. The time-series signal can then be analyzed for its frequency content as the Fourier coefficients of the transformed function which represents the contribution of each *sine* and *cosine* function at different frequency. The power density spectrum of the signal shows how the power of the periodic signal is distributed among various frequency components.

In a Fourier transform, the mapping from the time domain to the frequency domain, is done by means of the complex periodic plain wave functions of the form $y(f,t) = e^{\omega t}$ where $\omega = 2\pi f$.

The basic definition of a Fourier transform of a continuous time signal $x(t)$ is given by

$$X(\omega) = \int_{-\infty}^{\infty} x(t)e^{-i\omega t} dt \qquad (2.1)$$

As most of the continuous time-series data contains discrete-time data, the infinite integral can be replaced by finite summation as follows:

$$X(\omega_k) = \sum_{n=0}^{N-1} x(t_n)e^{-i\omega_k t_n}, n = 0,\ 1,\ 2,...,\ N-1 \qquad (2.2)$$

where N is the number of time samples and ω_k is the kth frequency. It is a Fourier representation of a finite length of sequence, which corresponds to samples equally spaced in frequency of the Fourier transform of the signal. For real valued signal $x(t)$, the Fourier transform is complex and in polar notation it can be represented as

$$X(\omega) = A(\omega)e^{i\theta(\omega)} \qquad (2.3)$$

where $A(\omega)$ is the spectral amplitude and $\theta(\omega)$ is the phase angle. These properties can be represented as

$$A(\omega) = \sqrt{R(\omega)^2 + I(\omega)^2} \qquad (2.4)$$

and

$$\theta(\omega) = \tan^{-1}\left(\frac{I(\omega)}{R(\omega)}\right) \qquad (2.5)$$

where $R(\omega)$ is the real (or *cosine*) part of the signal and $I(\omega)$ is the imaginary (or *sine*) part of the signal.

Fourier amplitude spectrum computes the harmonic components globally and thus yield average characteristics over the entire duration of the data. In order to investigate the time-frequency characteristics of nonstationary data, Fourier transform can be used with segments of the data to produce the representation called spectrogram and while using this each segments should be preferably stationary. However, when the length of window is shortened, the temporal resolution will be better while the quality of frequency resolution gets affected. Hence to achieve a trade-off between the two, an optimal window size is to be chosen in Fourier analysis, which is rather a challenging task. In short, although the Fourier transform provides useful information about a signal, it is often not enough to characterize signals whose frequency content changes in time. In such cases, the transforms that will enable us to obtain the frequency content of a process as a function of time is more appropriate. Such an analysis is called time-frequency analysis. The goal of time-frequency analysis is to expand a signal into waveforms whose time-frequency properties are adapted to the signal's local structure (Kumar and Georgiou 1997).

2.2.2 SHORT-TIME FOURIER TRANSFORM (STFT)

The Fourier transform converts a time domain signal to a frequency domain signal but offered some difficulties in analyzing nonstationary signals. To circumvent the limitations of Fourier transforms in analyzing the nonstationary signals locally, the Short Time Fourier Transform (STFT) was proposed which maps a signal into a two-dimensional function of time and frequency. In this method, the Fourier transform is used to analyze only a small section of the signal at a time (popularly known as windowing the signal), i.e., STFT break the nonstationary signals into sub signals each of which can be assumed to be stationary. Then the traditional FT can be applied, and energy density can be determined. This method provides some information about both when and at what frequencies a signal event occurs. Eventhough STFT evolved as a compromise between time and frequency information, it is observed that the information is obtained with limited precision, and that precision is highly depending on the size of the window chosen (Misiti et al. 2008). Further, the size for the time window is the same for all frequencies. To rectify this limitation, a more flexible approach with the use of variable window size, has become essential to get the information from practical signals with more accuracy in both time and frequency. The window function is user specified and the fixing of window sizing is a crucial step in STFT.

The mathematical form of STFT can be represented as follows:

$$STFT(\omega,t) = \frac{1}{2\pi} \int_{-\infty}^{\infty} x(t)h(t-\tau)e^{-i\omega t}d\tau \qquad (2.6)$$

In STFT, the plot of energy density contribution from each frequency is time dependent and it is popularly called as spectrograms.

2.3   ADVANCED SPECTRAL ANALYSIS METHODS

2.3.1   WAVELET TRANSFORM

Wavelet Transforms (WT) are popular tools processing geophysical signals (Kumar and Georgiou 1997). It is a scaling function which converts a signal into low frequency components and high frequency components.

Wavelets are the basic building blocks of WTs, which are small wave forms of limited duration and zero mean. The terminology regarding wavelet first appeared in 1909 in a thesis by Alfred Haar (1910), but a proper exploration of the theory of wavelets was done by researchers only in the second half of the 20th century.

Wavelet transforms are mathematical transforms that can be used to analyze time series that contain nonstationary power at many different frequencies (Daubechies 1990). Here, a variable-size windowing technique is adopted for time-frequency characterization. Wavelet transforms uses a wavelet prototype function called a mother wavelet. Initial forms of mother wavelets were proposed by Jean Morlet and team at the Marseille Theoretical Physics Center in France (Morlet et al. 1982). Later, different classes of wavelets were developed (Mallat 1989; Daubechies 1990; Coifman and Wickerhauser 1992; Kumar and Georgiou 1997). The wavelet transform is an inner product between a mother wavelet at a given scale and position with the signal to be analyzed. The scaled translation of mother wavelet function along time-series signal provides a series of wavelet coefficients based on the correlation between the two. In other words, the wavelet transform of a discrete data series is defined as the convolution between the data series with a scaled and translated version of the wavelet function chosen. Large values of the wavelet coefficients reflect the combined effects of a large fluctuation of the signal at this time and of a good matching of shape between the signal and the wavelet. The different steps involved in development of wavelet transform of a 1-D signal are given below:

1. Select a wavelet function (a mother wavelet) and compare it to a section at the beginning of the signal
2. Compute the coefficient of correlation between the original signal and the wavelet
3. Repeat steps (1) and (2) until the entire signal is covered
4. Stretch the wavelet and repeat steps (1) to (3)

Repeat the steps (1) to (3) for different levels (scales). The lower levels represent the compressed version of the wavelet and the higher levels are the stretched version of wavelet. The basic steps of wavelet analysis (shifting and stretching) are depicted in Figure 2.1.

Numerous mother wavelet functions have been proposed over the years (Sang et al. 2013) which differ in their mathematical form and character. The low frequency wavelets with long and high frequency wavelets with shorter base enable us to isolate very small events and large events easily. Mathematically, the wavelet transforms can also be designated as the weighted sum of the time series signal using the mother wavelets considering the infinite domain. There are two main types of wavelet transforms namely Continuous Wavelet Transform (CWT) and Discrete Wavelet

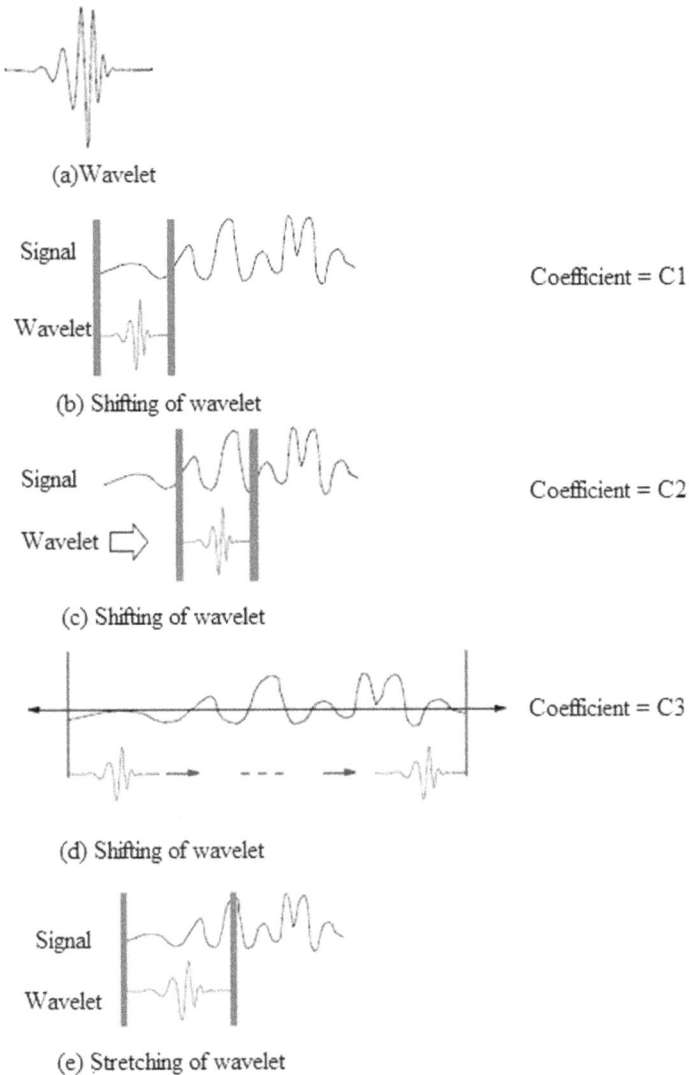

(a)Wavelet

(b) Shifting of wavelet

Signal

Wavelet

Coefficient = C1

(c) Shifting of wavelet

Signal

Wavelet

Coefficient = C2

(d) Shifting of wavelet

Coefficient = C3

(e) Stretching of wavelet

Signal

Wavelet

FIGURE 2.1 Illustration of basic steps of wavelet analysis.

Transform (DWT) that were being used for hydro-meteorological applications. The DWT is a special case of CWT in which a finite summation is followed. The CWT provides a flexible time-frequency window that automatically narrows when focusing on high-frequency oscillations and widens on the low-frequency background. A given time series signal is decomposed into specific number of high and low frequency components each with definite process scales. For characterization of hydro-meteorological datasets, the CWT is more suitable, while prediction problems the DWT is more appropriate because it decomposes the signal into a minimal number of independent coefficients.

2.3.1.1 Theory of Wavelet

Consider an input signal (time series), $X(t)t \in (-\infty, \infty)$. The wavelet function of time $\psi(t)$ is called mother wavelet, which satisfies (Burrus et al. 1998)

$$(i) \qquad \int_{-\infty}^{+\infty} \psi(t).dt = 0 \tag{2.7}$$

$$(ii) \qquad \int_{-\infty}^{+\infty} \left| \psi(t) \right|^2 dt < \infty \tag{2.8}$$

The property of the mother wavelet $\psi(t)$ is that the set of its continuous variants $\psi_{\tau,s}(t)$ forms an orthogonal basis in a two-dimensional space and it has finite energy and fast decay. Furthermore, the Fourier transform of the mother wavelet $\hat{\psi}(t)$ should satisfy admissibility conditions (Wei et al. 2012):

$$C_\omega = \int_R \left| \hat{\psi}(\omega) \right|^2 d\omega < \infty \tag{2.9}$$

$$\int_{-\infty}^{+\infty} \psi(t)dt = \hat{\psi}(0) = 0 \tag{2.10}$$

The wavelet transformation is the decomposition of $X(t)$ under different resolution levels (scales). The CWT of a discrete time-series signal X_i can be obtained from the convolution of $X_i(t)$ with the scaled and translated version of the mother wavelet function

$$W_X(\tau, s) = \sum_{i=1}^{N} X_i(t).\psi_{\tau,s}*(t) \tag{2.11}$$

where $\psi_{\tau,s}(t)$ is the normalized wavelet function and (*) represents the complex conjugate.

The normalization is done to ensure that wavelet transform at each scale is not weighted by the magnitude of the scale, which makes a direct comparison of different wavelet at different scales possible (Torrence and Compo 1998) and the normalization is done as

$$\psi_{\tau,s}(t) = \frac{1}{\sqrt{s}} \psi_{\tau,s}(t) \tag{2.12}$$

The stretching of the wave form $\psi(t)$ can be acquired through compressing and expanding it. The scaling factor for stretching or compressing the mother wavelet to the frequency of the signal is 's', and 'τ' is the translating factor (shifting factor) for shifting the mother wavelet to the time domain of the signal. The scaled and translated version of the mother wavelet is given by

$$\psi(t) = \psi\left(\frac{t-\tau}{s}\right) \qquad (2.13)$$

According to the mathematical definition of CWT, the wavelet analysis investigates the resemblance of the wavelet function with the signal in hand in the sense of frequency content, i.e., if a signal has a major component of frequency corresponding to the current scale, the wavelet at the current scale will be similar to the signal at the particular location where this frequency component occurs and thus the CWT coefficient at this point in time-frequency plane will be larger in magnitude. Thus, a spike is observed in the contour plot of CWT.

The computation of CWT coefficients can be made in a faster way, the operation is performed in a Fourier space, i.e., to get the CWT coefficient, first perform the Fourier transform of the signal and apply the convolution and subsequently do the inverse Fourier transform.

The choice of wavelet function is an important component in the wavelet transformation. Wavelet function can be a real or complex function. Complex wavelet functions make it possible to extract the information of both amplitude and phase, which is more suitable for analyzing the signal's oscillatory behavior (Kumar and Georgiou 1997; Torrence and Compo 1998). Morlet, Mexican Hat, and Haar are some of the mother wavelets usually employed in the CWT, out of which the Morlet is a complex function. The typical shapes of Mexican Hat, Morlet and Meyer mother wavelets are given in Figure 2.2.

The Morlet wavelet can be used to find the wavelet power spectra, which has found widespread application in hydrology due to its frequency resolution and ability to detect both time-dependent amplitude and phase for different frequencies in the time series (Torrence and Compo 1998; Labat 2005). The Morlet wavelet function can be defined by:

$$\psi_o(\eta) = \pi^{\frac{-1}{4}} e^{i\omega_o \eta} e^{\frac{-\eta^2}{2}} \qquad (2.14)$$

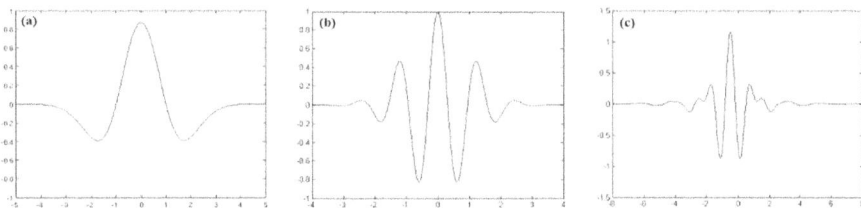

FIGURE 2.2 Examples of mother wavelet functions (a) Mexican Hat; (b) Morlet; and (c) Meyer.

where $\psi_o(\eta)$ is the wavelet function, η is a dimensionless parameter, i is the imaginary unit, ω_o is the dimensionless angular frequency, which provides a balance between time and frequency localaization.

For implementing WT, a set of scales are to be identified *apriori*. Here the scales are to be incremented continously to create a complete picture of WT. These set of sclaes can be generated using fractional powers of two (Torrence and Compo 1998).

$s_j = s_o 2^{j\delta j}$ *where j=0,1,....,J*, s_o is the smallest scale and J is the maximum number of scales to be investigated. δ_j is the scale step size whose value depends on the mother wavelet finction. For Morlet function δ_j, upto 0.5 can be chosen. The smallest scaleis $2\delta t$, where δt is the time step of the measured data.

The complex wavelet function results in complex wavelet coefficients constitute real (Re) and imaginary (Im) parts, i.e., amplitude $|W(\tau,s)|$ and phase given by $\tan^{-1}\left(\dfrac{Im(W(\tau,s))}{Re(W(\tau,s))}\right)$.

For convenient description, it is common to use the wavelet power spectrum $|W(\tau,s)|^2$ instead of the continous wavelet spectrum. The spectrum also normalized by its variance to enable the comparison among different wavelet spectra.

The calculation of wavelet coefficient at every possible scale generates a lot of data (which may sometime redundant also), which is computationally cumbersome. When the scaling factor and shifting factor of the basic wavelet function $\psi(a,\tau)$ are limited only to discrete values, the operation is termed as discrete wavelet transform (DWT). The details of wavelet transformation can be explained as follows:

$$\text{Let } a = a_0{}^j \text{ and } \tau = ka_0{}^j \tau_0 \text{ and } a_0 > 0 \text{ and } \tau_0 \in R, \quad \forall j,k = 0,1,2,....,m$$

$$\psi_{j,k}(t) = a_0{}^{-j/2}\psi(a_0{}^{-j}t - k\tau_0) \tag{2.15}$$

Then the wavelet transform can be expressed as:

$$W_\psi f(t) = a_0{}^{-j/2}\int_{-\infty}^{+\infty} f(t)\psi*(a_0{}^{-j}t - k\tau_0)dt \tag{2.16}$$

The simplest and most efficient practice is choosing $a_0 = 2$ and $\tau_0 = 1$; scale and position are selected based on the powers of base two logarithm, called *dyadic* scales and positions (Mallat 1989):

$$W_\psi f(j,k) = 2^{-j/2}\int_{-\infty}^{+\infty} f(t)\psi*(2^{-j}t - k)dt \tag{2.17}$$

In a discrete time series in which $f(t)$ occurs at discrete integer times steps, the dyadic DWT can be written as

$$W_\psi f(j,k) = 2^{-j/2} \sum_{j,k\in z} f(t)\psi^*(2^{-j}t-k)dt \tag{2.18}$$

where z is the domain of integer numbers.

Then the signal can be reconstructed using the following equation:

$$f(t) = \sum_{j,k\in z} W_\psi f(j,k)\psi_{j,k}(t) \tag{2.19}$$

As per the above equation, wavelet coefficients are divided into an approximation (or low frequency) coefficient (A_n) at level n through a low pass filter and detail (D) (or high frequency) coefficients at different levels $1, 2, ..., n$ through a high-pass filter. The approximation provides background information on the original signal, while $D_1, D_2, D_3, ..., D_n$ contains the detail information on the original signal (such as periodicity, break and jump). In a simplified form, the reconstruction stage can be expressed as:

$$f(t) = A_n(t) + \sum_{l=1}^{n} D_k(t) \tag{2.20}$$

where $A_n(t)$ is the approximation of the original signal at level n, and $D_l(t)$ are the details of the original signal at different levels ($l = 1, 2, 3,.., n$, the index for levels). $W_\psi f(a, \tau)$ and $W_\psi f(j,k)$ can reflect the characteristics of the original time series in frequency (a or j) and time domains(τ or k). When a or j is small, the frequency resolution is very low, but the time domain is very high. When a or j become large, the frequency resolution is high, but the time domain is low. The high and low frequency resolution of a signal S at single-level decomposition is shown in Figure 2.3 and a typical three-level decomposition is shown in Figure 2.4.

Wavelets transforms are used for analyzing the non-linear and non-stationary time series in which the data variations are intermittent or nonperiodic. But the selection of appropriate mother wavelet and appropriate decomposition levels are two major challenges faced by the modelers in using WT. Also, its efficacy to analyze time-frequency distributions was criticized since it has limited capabilities to handle the non-linearity of processes (Huang et al. 1998).

2.3.1.2 Types of Wavelets and Levels of Decomposition

The choice of an appropriate wavelet function is one of the most important step in time-series modeling using wavelets. The deterministic components (periodicities and trend) and noise in hydrologic data are generated by physical mechanisms and random factors. While applying the DWT upon hydrologic time series, the energy variation (i.e., the variance) of a series (known as energy function) of these components will be different. The selection of most suitable family should be based on the resemblance of the wavelet family with the time series data in hand. Many of the past studies followed a trial-and-error procedure for selection of a particular

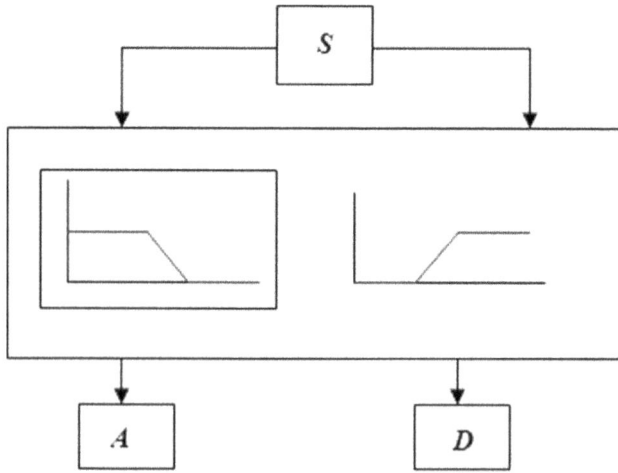

FIGURE 2.3   Single-level decomposition of a signal.

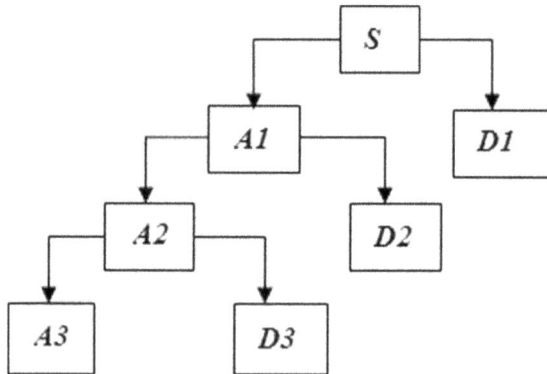

FIGURE 2.4   Three-level decomposition and reconstruction of a signal.

wavelet in wavelet-soft computing conjunction models for analyzing hydrologic time series data (Nourani et al. 2009). Recently some of the researchers gave more insight to this issue (Nourani et al. 2009; Maheswaran and Khosa 2012b; Sang 2013; Sang et al. 2016).To obtain good quality of compression and recovery during the procedure of signal compression, orthogonal wavelet base is usually being substituted for bi-orthogonal wavelet base (Chou 2007).

Brief descriptions of four families of wavelets, which are popular in hydrologic applications are given below:

   (a) **Haar Wavelet**: It is the simplest and first type of wavelets. It resembles a step function. It is a discontinuous type and having compact support (i.e., vanishes after a finite interval). Haar wavelet (Figure 2.5) has the advantage of orthogonality, but it has poor filtering characteristics (Chou 2007).

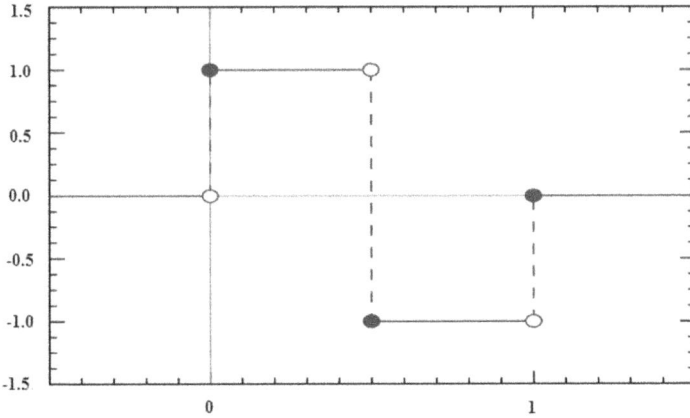

FIGURE 2.5   Haar wavelet.

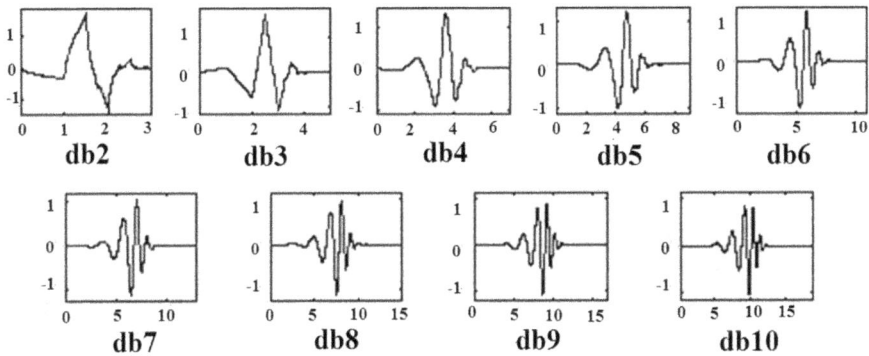

FIGURE 2.6   The Daubechies (db) family of wavelets.

The mother wavelet function can be represented as (Kumar and Georgiou 1997):

$$\psi(t) = \begin{cases} 1 & 0 \leq t < 0.5 \\ -1 & 0.5 \leq t < 1 \\ 0 & otherwise \end{cases} \tag{2.21}$$

(b) **Daubechies family**: Daubechies wavelets (Figure 2.6) are one of the most commonly used set of discrete wavelet transforms formulated by the Belgian mathematician Ingrid Daubechies in 1988 (Daubechies 1990). Daubechies wavelet family is usually written as 'dbN', where db is the 'surname', and N the order of the wavelet. According to Sang et al. (2013), the Daubechies family goes from 1 to 45. Daubechies No 1 (db1), the first and simplest one, represents the same Haar wavelet, whereas the complex ones are with many more oscillations in the tail.

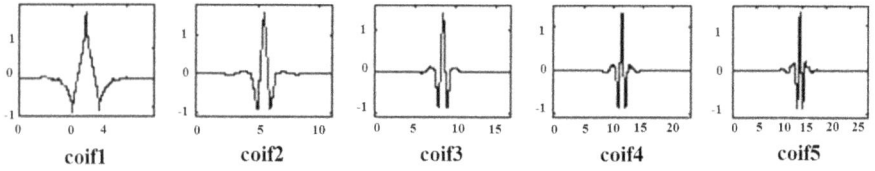

FIGURE 2.7   Coiflet family of wavelets.

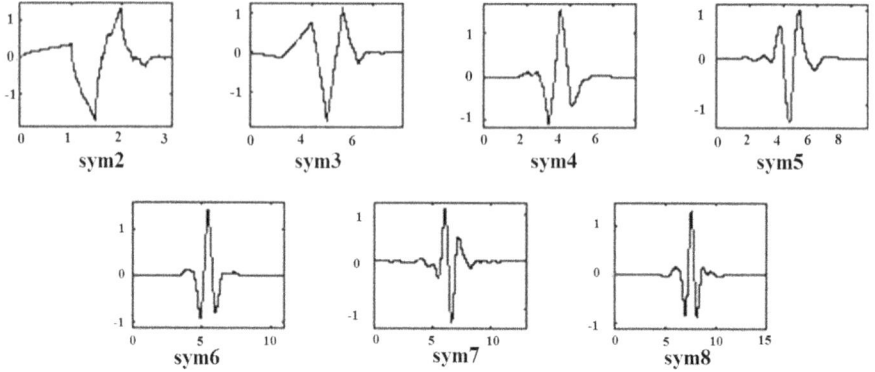

FIGURE 2.8   Symlet family of wavelets.

   (c) **Coiflet family**: Coiflet wavelets (Figure 2.7) are near symmetric wavelets
       having wavelet function with $2N$ vanishing moments and scaling functions
       with $2N$-1 vanishing moments, where N is the order of the wavelet. Both
       functions have a support of length $6N$-1.
   (d) **Symlet Family**: The symlets (Figure 2.8) are nearly symmetrical wavelets
       proposed by Daubechies as modifications to the db family. The properties of
       the two wavelet families are similar.

Sang et al. (2013) classified the entire wavelet families into three groups based on variation in energy function of noise and this classification is presented in Table 2.1. The
first family has 105 members, second have 11 members and third have 10 members.
According to Sang et al. (2013), the first group (group-I) that is favorable for
establishing the reference energy function is recommended for trend analysis studies.
The Haar, Coiflets (coif$N$, $N = 1, 2, …, 5$), Symlets (sym$N$, N=2, 3, …, 45), Daubechies
(db$N$, N=2, 3, …, 45), Biorsplines (bior1.1, bior1.3, bior1.5, bior4.4, bior6.8),
Reversebior (rbio1.1, rbio1.3, rbio1.5, rbio4.4, rbio6.8) belongs to the group-I and
these mother wavelets are popular among the hydrologists as the characteristics of
these wavelets matches well with most of the real field hydrologic time series data
(Nourani et al. 2011; Maheswaran and Khosa 2012a,b).

TABLE 2.1
Types of Wavelets

Varying rule of wavelet coefficient	Names of wavelet
First type (similar)	Haar (haar), Daubechies (dbN, N= 2,3,...,45), Symlets (symN, N= 2,3,. ...,45), Coiflets (coifN, N= 1,2,...,5), Bior Splines (bior1.1, bior1.3, bior1.5, bior4.4, bior6.8), Reverse Bior (rbio1.1, rbio1.3, rbio1.5, rbio4.4, rbio6.8)
Second type (decreasing)	Dmeyer (dmey), Bior Splines (bior5.5), Reverse Bior (rbio2.2, rbio2.4, rbio2.6, rbio2.8, rbio3.1, rbio3.3, rbio3.5, rbio3.7, rbio3.9)
Third type (increasing)	Bior Splines (bior2.2, bior2.4, bior2.6, bior2.8, bior3.1, bior3.3, bior3.5, bior3.7, bior3.9), Reverse Bior (rbio5.5)

Source: Sang et al. (2013).

2.3.1.3   Levels of Decomposition

The selection of optimum number of decomposition levels in a particular modeling problem is a challenging task. Most of the studies followed a trial and error procedure and some studies followed empirical formulae for fixing the number of decomposition levels.

1. Aussem et al. (1998) proposed $L = $ int $\{\log (N)\}$ for fixing the number of decomposition levels. Where N is the data length and 'log'; stands for base 10 logarithm. The criteria is widely accepted by researchers (Wang and Ding 2003; Kisi 2009; Nourani et al. 2009).
2. Sang et al. (2013) suggested that $\log_2(N)$ can be used as a criteria for trend analysis studies.
3. Misiti et al. (2008) suggested that the maximum allowed level of decomposition as 5 for the one-dimensional wavelet decomposition.

2.3.1.4   Cross Wavelet Spectrum

Using cross wavelet spectrum, cyclic features at which the underlying time series are covarying can be detected. The covariations of two signals demonstrate the existence of a link between the underlying processes and also the fact that the information of one process is capable of predicting the other process. This information is very useful when it is of interest to find out the processes that have correlation (or strong correlation) with a target time series. The signals, which are showing to have high common power with the target signal in the cross wavelet spectrum, can be used as predictors in the estimation of temporal variations of the target time series. This is important information in the modeling of complex processes, such as hydrometeorological processes, in which determination of important predictors (input variables) is essential and a challenging task. The cross wavelet transform (XWT) of two time series

X_n and Y_n is defined as $W_n^{XY} = W_n^X.W_n^{Y*}$, where * denotes complex conjugation. The cross wavelet power can be further defined as $|W^{XY}|$. The local relative phase between X_n and Y_n in time frequency space can be obtained from the $\arg(W^{XY})$. The theoretical distribution of the cross wavelet power of two time series with background power spectra P_k^X and P_k^Y is given by (Torrence and Compo 1998):

$$D\left(\frac{|W_n^X(s)W_n^{Y*}(s)|}{\sigma_X\sigma_Y} < p\right) = \frac{Z_v(p)}{v}\sqrt{P_k^X P_k^Y} \qquad (2.22)$$

where k is the Fourier frequency index, probability that a wavlet power of a process can be greater than p; $v = 1$ and 2, respectively, for real and complex wavelets.

From the two CWTs the XWT is calculated. The XWT exposes regions with high common power and further reveals information about the phase relationship. If the two series are physically related, one can expect a consistent or slowly varying phase lag (Grinsted et al. 2004). The circular mean of the phase angles can be used to quantify the phase relationship.

2.3.1.5 Wavelet Coherence (WTC)

Wavelet coherence is defined as the square of the cross-spectrum normalized by the individual power spectra. It is the localized correlation coefficient between the two time series in the time-frequency domain. Consider two time series X and Y having wavelet coefficients $W_n^X(s)$ and $W_n^Y(s)$, where n is the localized time index and s is the scale:

$$R_n^2(s) = \frac{|S(s^{-1}W_n^{XY}(s))|^2}{S(s^{-1}W_n^X(s)) \times S(s^{-1}W_n^Y(s))} \qquad (2.23)$$

In the above expression, S is the smoothing operator in both time and frequency domain given by $S(W) = S_{scale}(S_{time}(W_n(s)))$, where S_{scale} smoothens the wavelet transform in the frequency axis and S_{time} smoothens it along the time axis. The value varies from 0 to 1, 1 indicates high correlation and zero indicate low correlation.

2.3.2 HILBERT-HUANG TRANSFORM

In the recent times HHT is gaining huge popularity for spectral analysis of non-linear and nonstationary time series data (Huang et al. 2016). It is a purely data adaptive method, which can produce physically meaningful representations of time-series data. This method does not require *a priori* selection of functions, but instead it decomposes the signal into intrinsic oscillation modes derived from the succession of extrema. The HHT involves two major steps (i) the use of EMD method or its variants such as EEMD (Wu and Huang 2005) or CEEMDAN (Torres et al. 2011) to decompose a time series into a collection of orthogonal time series, namely IMFs; (ii) the use of HSA to obtain instantaneous frequency, which may be helpful to identify embedded structures of time series data. This decomposition helps to better understand the internal structure of the signal and the components involved (Klionski et al.

2008). The original data can be reconstructed by the linear addition of each of the IMFs, which are independent to each other.

Each IMF must meet the following conditions:

(a) the number of extreme values in the overall data must match the number of zero crossings, or differ by only one

$$N_{max} + N_{min} = N_{zero} \pm 1 \qquad (2.24)$$

where N_{max} = total number of maxima; N_{min} = total number of minima; and N_{zero} = total number of zero crossings; and

(b) at any point of time, the mean value of the envelope of local maxima and envelope of local minima ($m(t)$), must be zero

$$m(t) = \frac{E_{max}(t) + E_{min}(t)}{2} \approx 0 \qquad (2.25)$$

where $E_{max}(t)$ is the envelope of the local maxima; and $E_{min}(t)$ is the envelope of the local minima, which were obtained by spline interpolation procedures.

2.3.2.1 Empirical Mode Decomposition

The process of extracting the IMFs from a time series $X(t)$ (which is known as the sifting process) consists of the following steps:

1. Identify all extrema (maxima and minima) of the signal $X(t)$
2. Connect these maxima with any interpolation function (e.g., cubic spline) to construct an upper envelope ($E_{max}(t)$); use the same procedure for minima to construct a lower envelope ($E_{min}(t)$)
3. Compute the mean of the upper and lower envelope, $m(t)$
4. Calculate the difference time series $d(t) = X(t)-m(t)$
5. Let $d(t)$ be the new signal and repeat steps (1) to (4) until $d(t)$ becomes a zero-mean series with no riding waves (i.e., there are no negative local maxima and positive local minima) with smoothened amplitudes.

To satisfy step (5), an appropriate criterion is to be applied to stop the sifting iterations in order to guarantee that the IMF retains enough physical sense of both amplitude and frequency modulations (Huang and Wu 2008). A number of stopping criteria have been reported in the literature (Huang et al. 1998; Huang and Wu 2008). One popular criteria is the modified Cauchy type stopping criterion (Huang and Wu 2008) computed from two consecutive sifting results in:

$$\frac{\sum_{t=0}^{T} \left| d_{ki-1}(t) - d_{ki}(t) \right|^2}{\sum_{t=0}^{T} \left| d_{ki-1}(t) \right|^2} \leq \xi \qquad (2.26)$$

where 'k' is the index for IMF; 'i' is the index for iteration of the sifting operation (to get k^{th} IMF); T is the data length; $d_{k,i}(t)$ is the deviation of the original time series from the mean in the ith iteration to evolve the kth IMF; 'ξ' is the tolerance value specified by the user (normally, 0.2–0.3 as per Huang and Wu 2008).

6. On satisfying the zero-mean condition, $d(t)$ can be designated as the first intrinsic mode function IMF1.
7. Compute the residue $R_1(t)$ by subtracting IMF1 from original signal (i.e., $R_1(t)=X(t)-IMF_1(t)$), is used as new signal. The sifting process is repeated upon this signal to get IMF2.
8. The higher oscillatory modes are obtained by treating the residue ($R_k(t)$) as the signal ($X(t)$), iteratively.

The kth residue is defined as $R_k(t) = X(t) - \sum_{j=1}^{k} IMF_j(t)$. The process will be continued till the resulting residue is a monotonic function or a function having only one extrema.

Then the original signal can be reconstructed as $X(t) = \sum_{k=1}^{K} \left[IMF_k(t) \right] + R_K(t)$, where K is the number of decomposed IMFs.

During the decomposition of a signal into IMFs, sometimes it may fail to assign the signals with similar frequencies into separate IMFs. As a result, each IMF contains different modes of oscillations, which make IMFs to lose physical meaning or falsely represent the physical processes associated with mode. To solve this problem, noise assisted data analysis methods were proposed by different researchers in the past. The Ensemble EMD (EEMD) (Wu and Huang 2005) and the Complete Ensemble EMD with Adaptive Noise (CEEMDAN) (Torres et al. 2011) falls in this category.

2.3.2.2   Ensemble Empirical Mode Decomposition
The procedure of EEMD is presented below (Wu and Huang 2005):

1. Initialize the number of ensemble realizations M, the amplitude of the added white noise for the mth realization (w_m), and set $m = 1$
2. Perform m^{th} trial on the signal
    (i) Add white noise series with the given amplitude to the investigated signal

$$X_m(t) = X(t) + w_m(t)$$

   where $X_m(t)$ indicates the mth artificial signal; $X(t)$ is the true signal; and $w_m(t)$ represents the zero mean unit variance white noise signal for the mth trial.
    (ii) Decompose the noise-added signal $X_m(t)$ into K number of IMFs $IMF_{k,m}$ ($k = 1,2,..., K$) using the EMD method described in the previous section, where $IMF_{k,m} = k^{th}$ IMF of the mth trial

(iii) If $m < M$, go to step (i) with $m = m + 1$. Repeat Steps (i) and (ii) multiple times, with different white noise series

3. Obtain an ensemble mean $\overline{IMF_k(t)}$ of the corresponding IMFs; report the mean $\overline{IMF_k(t)}$ as the final IMF.

The effect of the added white noise should decrease following statistical rules (Wu and Huang 2005):

$$\psi_f = \frac{\psi}{\sqrt{N}} \tag{2.27}$$

$$\ln \psi_f + \frac{\varepsilon}{2}\ln N = 0 \tag{2.28}$$

where ψ is the amplitude of the added noise and ψ_f is the final standard deviation of error, which is defined as the difference between the input signal and the corresponding IMFs. Equations (2.27) and (2.28) indicate that as the ensemble number increases, the effect of added white noise gets nullified. In short, k^{th} true IMF $(IMF_k(t))$ is the mean of the corresponding IMFs obtained by EMD over ensemble of artificial signals, generated by adding different realizations of white noise $(w_m(t))$ to the original signal $X(t)$.

2.3.2.3 Complete Ensemble Empirical Mode Decomposition with Adaptive Noise (CEEMDAN)

In CEEMDAN method, a noise is added at each stage of the decomposition to result in a unique residue in each mode from the residue of previous mode (or the true signal, for the first mode) and currently generated IMF.

1. Perform the EMD for M realizations $X_m(t) = X(t) + \varepsilon_0 w_m(t)$ and compute the first mode of CEEMDAN $\overline{IMF_1(t)}$, by averaging the realizations

$$\overline{IMF_1(t)} = \frac{1}{M}\sum_{m=1}^{M} IMF_m(t) \tag{2.29}$$

where $m=1,2,...,M$ the index for realizations; ε_o is the noise parameter for the initial step.

This step shows that the first mode obtained by CEEMDAN will be same as that obtained by EEMD.

2. At the first stage $(k = 1)$ calculate the first residue as $R_1(t) = X(t) - \overline{IMF_1(t)}$
3. Decompose the realizations $R_{1m}(t) = R_1(t) + \varepsilon_1 E_1(w_m(t))$, $m = 1,2,...,M$ until their first EMD mode gets evolved. Then compute the second mode $\overline{IMF_2(t)}$

$$\overline{\overline{IMF_2(t)}} = \frac{1}{M} \sum_{m=1}^{M} E_1[R_1(t) + \varepsilon_1 E_1(w_m(t))] \tag{2.30}$$

where the operator $E_k(.)$ is an operator represents the evolution of the kth mode by EMD; ε_1 is the noise parameter for the first stage ($k = 1$)

4. Calculate the kth residue as

$$R_k(t) = R_{k-1}(t) - \overline{\overline{IMF_k(t)}} \text{ For k=2, 3,..., K} \tag{2.31}$$

where $\overline{\overline{IMF_k(t)}}$ is the IMFs obtained by CEEMDAN

5. Compute the first mode of $R_k(t) + \varepsilon_k E_k(w_m(t))$, and define the $(k+1)^{th}$ mode by CEEMDAN as

$$\overline{\overline{IMF_{k+1}(t)}} = \frac{1}{M} \sum_{m=1}^{M} E_1[R_k(t) + \varepsilon_k E_k(w_m(t))] \tag{2.32}$$

6. Go to step (4) for next k

Steps 4 to 6 are performed until the obtained residue is no longer feasible to be decomposed (i.e., the residue is monotonic or having only one extrema).

The final residue becomes $R_K(t) = X(t) - \sum_{k=1}^{K} \overline{\overline{IMF_k(t)}}$. Therefore, the above decomposition process is complete and provides an exact reconstruction of the original data, i.e.,

$$X(t) = \sum_{k=1}^{K} \left[\overline{\overline{IMF_k(t)}} \right] + R_K(t) \tag{2.33}$$

Most of the geophysical processes are influenced by multiple potential causal variables and it is often necessary to investigate the association of processes with such input (causal) variables. Over the years, researchers developed different variants of EMD (or its noise-assisted versions), which can handle multiple time series simultaneously. For example, Rilling et al., (2007) proposed bidimensional EMD, Rehman and Mandic (2010a) proposed tridimensional EMD, etc. The Multivariate EMD (MEMD) proposed by Rehman and Mandic (2010b) is the most generalized version of EMD to decompose multiple time series signals simultaneously.

2.3.2.4   Multivariate Empirical Mode Decomposition

Multivariate EMD proposed by Rehman and Mandic (2010b) is an extension of the traditional EMD, which decomposes multiple time series simultaneously after identifying the common scales inherent in different time series of concern. In this method, multiple envelops are produced by taking projections of multiple inputs along different directions in an m-dimensional space.

Assuming $V(t) = \{v_1(t), v_2(t) \dots v_m(t)\}$ being the m vectors as a function of time t, and $X^{\varphi_k} = \{x_1^k, x_2^k, \dots, x_m^k\}$ denoting the direction vector along different directions given by angles $\varphi_k = \{\varphi_1^k, \varphi_2^k, \dots \varphi_{m-1}^k\}$ in a direction set X ($k = 1,2,3, \dots K$, K is the total number of directions). It can be noted that the rotational modes appear as the counterparts of the oscillatory modes in EMD or its variants. The IMFs of m temporal datasets can be obtained by the following steps:

1. Generate a suitable set of direction vectors by sampling on a ($m - 1$) unit hypersphere
2. Calculate the projection $p^{\varphi_k}(t)$ of the datasets $V(t)$ along the direction vector X^{φ_k} for all k
3. Find temporal instants $t_i^{\varphi_k}$ corresponding to the maxima of projection for all k
4. Interpolate $[t_i^{\varphi_k}, V(t_i^{\varphi_k})]$ to obtain multivariate envelop curves $e^{\varphi_k(t)}$ for all k
5. The mean of envelope curves ($M(t)$) is calculated by $M(t) = \dfrac{1}{K} \displaystyle\sum_{k=1}^{K} e^{\varphi_k}(t)$
6. Extract the detail $D(t)$ using $D(t) = V(t) - M(t)$. If $D(t)$ fulfills the stoppage criterion for a multivariate IMF, apply the above procedure from step (1) onward upon the residue series (i.e., $V(t) - D(t)$). Otherwise, repeat the steps from (2) onward upon the series $D(t)$.

Hammersley sampling sequence can be used for the generation of direction vectors (Huang et al. 2016) and the stoppage criteria proposed for EMD (i.e., the sum squared difference between the deviations of the mode from the mean signal in two consecutive iterations should be less than a specified tolerance) can be used in the implementation of MEMD.

Multivariate dataset could be a collection of totally different signals and MEMD is a technique, which can capture the common scales present in those signals. The extension of EMD to a multivariate dataset involves the following steps: (i) performs multiple real-valued projections of the signals along multiple directions on hyperspheres (n-spheres); (ii) multidimensional envelopes of the signal are created by interpolating (component wise) the extrema of projected signals; (iii) averaging of the envelops to obtain the local mean. In the implementation of the above steps, MEMD uses the mathematical concept of quaternion. The quaternions, first described by Irish mathematician Sir William Rowan Hamilton in 1843, are a number system that extends the complex numbers to higher dimensions.

Any number of the form $q = a + bi + cj + dk$, where a, b, c, and d are real numbers, $i^2 = j^2 = k^2 = -1$, and $ij = k = -ji$, $jk = i$ and $ki = j$; is a quaternion.

Under addition and multiplication, quaternions have all the properties of a field, except multiplication is not commutative.

The norm of quaternions is defined as $\|q\| = \sqrt{a^2 + b^2 + c^2 + d^2}$ and the unit vector is $e^{u\theta} = \cos\theta + u\sin\theta$.

The conceptual idea of projection of direction vectors and sampling are provided in Figures 2.9 and 2.10.

To generate a suitable set of direction vectors, unit hyperspheres can be sampled based on uniform angular sampling and quasi–Monte Carlo-based low-discrepancy

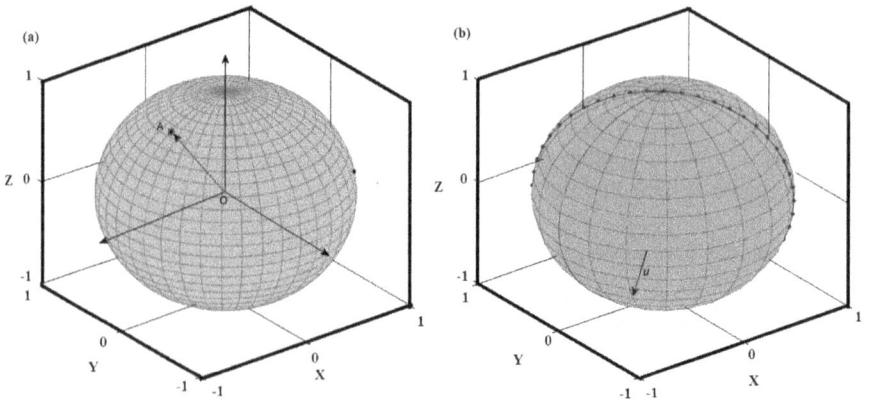

FIGURE 2.9 Concept of projection of a point along different directions in a 3D space: (a) the direction vector OA in 3D space, can also be represented by a point on the surface of a unit sphere; (b) multiple direction vectors represented by points on a particular longitude line.

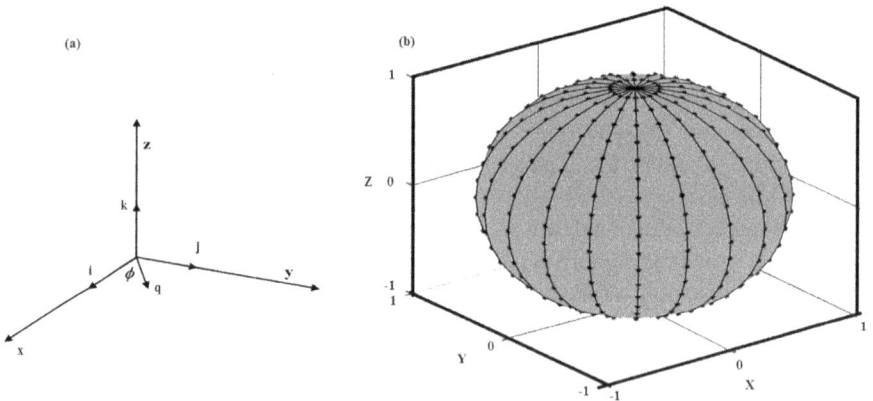

FIGURE 2.10 Concept of projection of a signal along multiple directions by rotation (a) For projections along longitudinal lines on a sphere, multiple axes represented by a set of vectors are chosen in the x–y plane, with angle taken with respect to x-axis; (b) projections of the input signal can be taken by rotating the input signal along rotation axes represented by a set of unit quaternions.

sequences. A simple and practically convenient choice for a set of direction vectors is to employ uniform angular sampling of a unit sphere in an n-dimensional hyperspherical coordinate system. The resulting set of direction vectors covers the whole $(n-1)$ *sphere*. A coordinate system in an n-dimensional Euclidean space can then be defined to serve as a point set (and the corresponding set of direction vectors) on an $(n-1)$ *sphere*. Let $\{\theta_1, \theta_2, \dots, \theta_{(n-1)}\}$ be the $(n-1)$ angular coordinates, then an

n-dimensional coordinate system having $\{x_i\}_{i=1}^{n}$ as the n coordinates on a unit $(n-1)$ sphere is given by:

$$x_1 = \cos(\theta_1)$$

$$x_2 = \sin(\theta_1)\cos(\theta_2)$$

$\cdot$

$\cdot$

$\cdot$

$$x_3 = \sin(\theta_1)\cos(\theta_2)\cos(\theta_3)$$

$$x_{n-1} = \sin(\theta_1)\ldots\ldots\ldots\ldots\sin(\theta_{n-2})\cos(\theta_{n-1})$$

$$x_n = \sin(\theta_1)\ldots\ldots\ldots\ldots\sin(\theta_{n-2})\sin(\theta_{n-1})$$

The point set corresponding to the n-dimensional coordinate system is now very convenient to generate. There are many ways to generate such sequence. In certain applications the quasi–Monte Carlo method and their lower discrepancy (close to uniform sampling) version is used. Rehman and Mandic (2010b) used Hammersley-Halton sequence (involving prime numbers, quasi–Monte Carlo method) to generate the direction vectors. Multiple direction vectors in n-dimensional space is depicted in Figure 2.11.

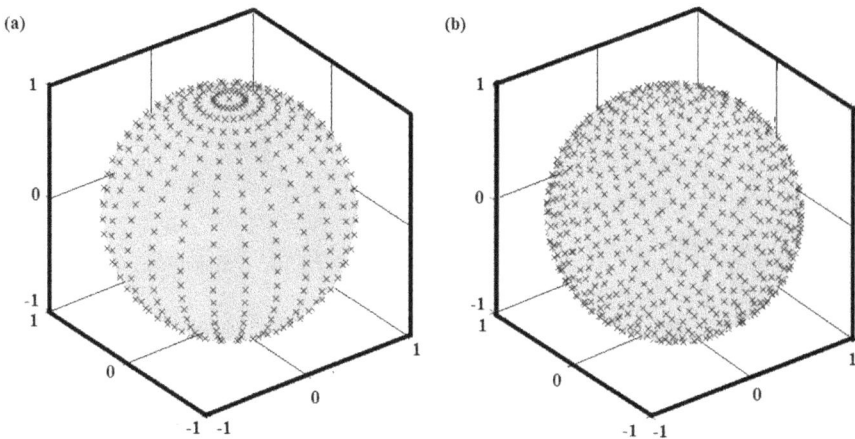

FIGURE 2.11   Direction vectors for taking projections of tri-variate signals on a two-sphere generated by using (a) spherical coordinate system and (b) a low-discrepancy Hammersley sequence.

2.3.2.5 Statistical Significance Test of IMF Components

The expected energy distribution of IMFs (squared sum of the IMFs averaged over the data length) is not enough to tell whether an IMF of real data contains only noise components or not, because the energy distribution of IMFs of a noise can deviate significantly from expected energy distribution. Therefore, the spread of energy distribution of noise is also to be estimated to identify whether a signal contain information or not (Huang and Wu 2008). For this purpose, Wu and Huang (2004a, 2004b) proposed a statistical significance test for IMF components. Eventhough the presented significance test suffers from the drawbacks of subjective nature in selection of noiseless components and inability to assess the presence of red noise (Lee and Ouarda 2011), the method is still popular and practical approach for identifying the significant components in a time series.

The procedure for statistical significance test of IMF components is given below:

1. Calculate the energy density (E_k) of each IMF as:

$$E_k = \frac{1}{N} \sum_{j=1}^{N} IMF_k^2(j) \tag{2.34}$$

   where k is the index for IMF, and N is the data length.
2. Select any one IMF as reference IMF composed mainly of noise and use the energy density of the reference IMF to normalize the energy density of rest of the IMFs, i.e., divide each energy density by the energy density of reference IMF. Generally, the first IMF is assumed to be composed of random noise and selected as the reference IMF (Wu and Huang 2004b)
3. Generate a white noise data by Monte Carlo method and divide it into sections of equal length of the series as that of $X(t)$. Apply EMD to decompose each section of white noise data into IMFs and calculate the spread function of white noise as:

$$\rho(y_k) = C' \exp\left\{ -\frac{N\overline{E}(y_k - \overline{y})}{4}^2 \right\} \tag{2.35}$$

   where $C' = N^{\overline{NE}\backslash2} \exp\left(-\frac{N\overline{E}(1-\overline{y})}{2} \right)$, $y_k = \ln(E_k)$, $\overline{y}$ and $\overline{E}$ are the mean of y_k and E_k, respectively.
4. Estimate the confidence line of the white noise corresponding to a selected confidence level (say 95 %) by

   $y_k = -x_k \pm Z\sqrt{\frac{2}{N}} \exp^{x_k\backslash2}$ where $x_k = \ln(\overline{T}_k)$ and $\overline{T}_k$ is the mean period calculated from the spectrum of IMF_k; Z is the constant estimated for the selected confidence level from the standard normal distribution table.

5. Plot the upper confidence line and mark the points $(\log_2(MNE_k), \log_2(MP_k))$ for different IMFs $(k=1,2,...,K)$; MNE_k is the mean normalized energy of kth IMF and MP_k is the mean period of kth IMF.
6. Compare the energy level of different IMFs with the spread function of white noise. The IMFs of series $X(t)$ that have their energy densities located above the upper confidence line of the white noise series can be considered to be a true IMF (free from white noise and contain information) and statistically significant at the selected confidence level.

Wu and Huang (2004a) suggested that the first IMF is not considered while making conclusions based on the significance test as it contains no perceivable physical process. Also the last IMF (residue) often has much higher energy level than other IMFs (Huang et al. 1998) and hence it will be above the upper confidence level in most of the cases.

2.3.2.6 Hilbert Transform and Its Normalization-Direct Quadrature Scheme

In the second phase of HHT, the IMFs of the time-series signal $X(t)$ obtained in the first phase (say $IMF(t)$) should be subjected to Hilbert Transform (HT), to yield instantaneous frequency and instantaneous amplitudes. Hilbert transform is the convolution of $IMF(t)$ with the function $g(t) = \dfrac{1}{\pi t}$. As $g(t)$ is nonintegrable, the HT of $IMF(t)$ is presented with the Cauchy Principal Value (Kanwal 1996) as:

$$Y(t) = H[IMF(t)] = PV \int_{-\infty}^{\infty} IMF(\tau)g(t-\tau)d\tau$$

$$= \frac{1}{\pi} PV \int_{-\infty}^{\infty} \frac{IMF(\tau)}{t-\tau} d\tau = -\frac{1}{\pi} \lim_{\tau \to 0} \int_{\tau}^{\infty} \frac{IMF(t+\tau) - IMF(t-\tau)}{\tau} d\tau$$

where PpV is the Cauchy principal value. Hence, an analytic signal $(Z(t))$ can be represented by combining $IMF(t)$ and $Y(t)$ as follows:

$$Z(t) = IMF(t) + iY(t) = A(t)e^{i\theta(t)} \tag{2.37}$$

where $i = \sqrt{-1}$; $A(t)$ is the amplitude, $\theta(t)$ is the phase angle, which are defined as:

$$A(t) = \sqrt{IMF^2(t) + Y^2(t)} \tag{2.38}$$

$$\theta(t) = \frac{Y(t)}{IMF(t)} \tag{2.39}$$

The instantaneous frequency is given by

$$\omega(t) = \frac{d\theta(t)}{dt} \tag{2.40}$$

The instantaneous frequency is fully capable of describing not only inter-wave frequency changes due to nonstationarity but also the intrawave frequency modulation due to nonlinearity (Huang and Wu 2008). Thus, HHT can distribute the amplitudes on the time-frequency plane. The instantaneous frequency of each IMF can be detected from the Hilbert spectrum. Therefore, HHT is a better tool for time-frequency analysis of nonstationary signals particularly when they carry non-linear features of a physical phenomenon.

If the instantaneous frequencies and instantaneous amplitudes of IMFs are obtained by Hilbert Transformation of IMFs, the time series $X(t)$ can be expressed as

$$X(t) = \mathrm{Re}\left[\sum_{k=1}^{K} A_k(t) e^{i\int \omega_k(t)dt} \right] + R_K(t) \tag{2.41}$$

The Fourier representation of the same data is

$$X(t) = \mathrm{Re}\left[\sum_{k=1}^{K} A_k e^{i\omega_k t} \right] \tag{2.42}$$

where Re[] stands for the real part and $(k = 1,2,\ldots, K)$ is the index for modes.

The time-frequency distribution of the amplitude is designated as the Hilbert amplitude spectrum (or Hilbert spectrum)$H(\omega,t)$, which can be defined as (Rudi et al. 2010):

$$H(\omega,t) = H[\omega(t),t] = \left\{ A_i(t) \text{ on the curve } \left\{ [\omega(t),t] : t \in R \right\} \right. \tag{2.43}$$

where $I = 1,2,\ldots, N$ is index of IMFs.

Thus, HHT can distribute the amplitudes on the time-frequency plane and the Hilbert energy spectrum can be defined by considering second-order amplitudes as follows:

$$H(\omega,t) = A^2(\omega,t) \tag{2.44}$$

The Hilbert spectrum is representing the energy-time-frequency information at local level and its integration over the time gives the marginal Hilbert spectrum:

$$h(\omega) = \int_0^\infty H(\omega,t).dt \tag{2.45}$$

An alternative way to define the marginal Hilbert spectrum is to define the joint pdf $p(\omega, A)$ of instantaneous frequency ω and amplitude A (Huang et al. 2008). Then the marginal Hilbert spectrum can also be written as the second-order statistical moment:

$$h(\omega) = \int_0^\infty p(\omega, A) A^2 dA \tag{2.46}$$

The above equation can be expressed as a generalized one and by defining arbitrary order moments, Huang (2009) proposed the Arbitrary Order Hilbert Spectral Analysis (AOHSA). According to which, the marginal spectrum is expressed by:

$$L_q(\omega) = \int_0^\infty p(\omega, A) A^q dA \text{ where } q \geq 0 \tag{2.47}$$

The AOHSA method can be used to detect the multifractality of the time series. In this process, the Hilbert marginal spectrum is constructed for different q orders. Theoretically, Huang (2009) defined the spectrum for $q = -1$ onward, but for streamflow analysis it is recommended to select q in the range 0 to 5 (Pande et al. 1998; Huang et al. 2009a). The arbitrary order spectra can be expressed in such a way that it follows scaling law within certain scale range.

$L_q(\omega) = \omega^{-\xi(q)}$ where $\xi(q)$ is the scaling exponent in the Hilbert space, i.e., the slope of the log-log plot of Hilbert spectra within the chosen scale range corresponding to different q order moments gives the scaling exponent. If the plot of scaling exponents (corresponding to different q-orders) follows a concave shape it can be concluded that the series is mulifractal in the range chosen. The scaling exponent $\xi(q)$ can be linked to scaling exponent $\zeta(q)$ of structure functions (Huang et al. 2008)as follows:

$$\xi(q) = \zeta(q) + 1 \tag{2.48}$$

Furthermore, $\xi(q)$ the scaling exponent can be related to the classical Hurst exponent (H) by the following relation:

$$\xi(q) = qH + 1, \text{ where } H = \xi(1) - 1. \tag{2.49}$$

For an uncorrelated series, the value of Hurst exponent is 0.5. If the Hurst exponent falls between 0.5 and 1, it indicates the long-term persistence (long memory process) and if it falls between 0 and 0.5 indicate a short term persistence (short memory process). If the value of H is one, it indicates a trending series. If the scaling exponent $\xi(q)$ for $q = 1$ exceeds 2, the relation between the fractal dimension (D) and H fails, as the fractal dimension generally defined upto 2. In such cases the second order relation $\xi(2) = 2H + 1$ can be used to estimate the value of Hurst exponent, H. More mathematical details can be found in Huang (2009) and proof for the above relations can be found in Zhou et al. (2013).

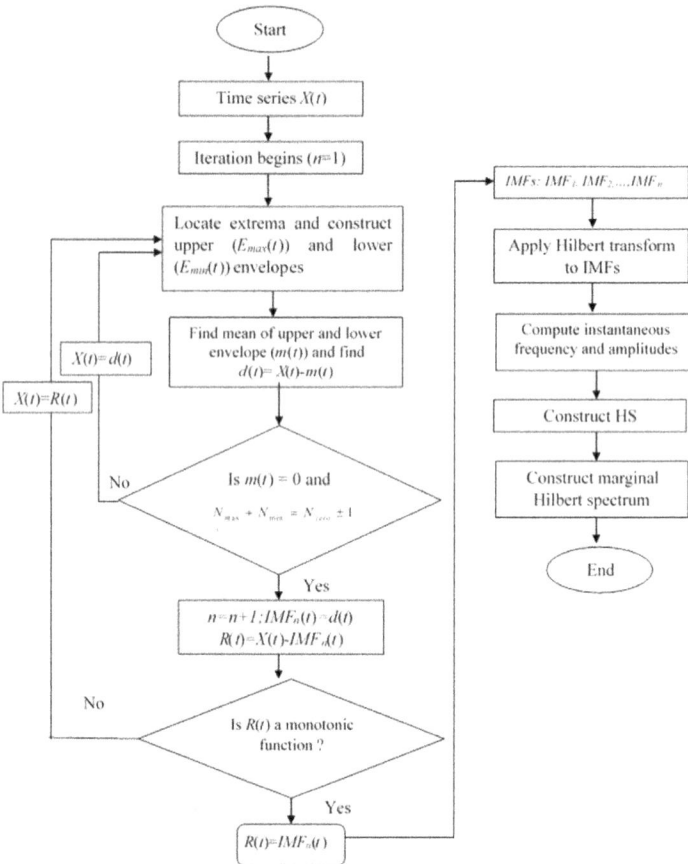

FIGURE 2.12 Flowchart of Hilbert-Huang transform.

The complete HHT procedure is depicted in Figure 2.12.

Hilbert transform has been considered unusable or imprecise in defining the instantaneous frequency by two well-known theorems, namely the Bedrosian theorem (Bedrosian 1963) and the Nuttall theorem (Nuttall 1966). On applying the Hilbert Transform, for any IMF $x(t) = A(t)\cos\theta(t)$, to yield physically meaningful instantaneous frequency (such as negative frequency), the following condition is to be satisfied:

$$H[A(t)\cos\theta(t)] = A(t)H[\cos\theta(t)] \tag{2.50}$$

But according to Bedrosian theorem (1963), the Hilbert transform for the product of two functions, $f(t)$ and $h(t)$ can be written as $H[f(t)h(t)] = f(t)H[h(t)]$ only if the Fourier spectra for $f(t)$ and $h(t)$ are totally disjoint in frequency space and the frequency range of the spectrum for $h(t)$ is higher than that of $f(t)$.

Thus, Equation (2.26) can be true only if the amplitude is varying so slowly that the frequency spectra of the envelope and the carrier waves are disjoint. This condition is seldom satisfied for practical datasets. To avoid this difficulty, Huang et al. (2009b) proposed a normalization scheme, which is an empirical method to separate the IMF into amplitude modulation (AM) and frequency modulation (FM) parts uniquely. After separation, the Hilbert transform is applied on the FM part alone to avoid the difficulty stated in the Bedrosian theorem (1963) and Nuttall theorem (1966). This guarantees that the Hilbert transform of $A(t)\cos\theta(t)$ is given by $A(t)\sin\theta(t)$. Nuttall theorem (Nuttall 1966) further stipulates that the Hilbert transform of $\cos\theta(t)$ is not necessarily $\sin\theta(t)$ for an arbitrary function $\theta(t)$. In other words, there is a discrepancy between the Hilbert transform and the perfect quadrature of an arbitrary function $\theta(t)$. The restriction of the Bedrosian theorem was surpassed through the EMD process and normalization of the resulting IMFs and the restriction imposed by Nuttall theorem was circumvented by the direct quadrature method proposed by Huang et al. (2009b).

The steps involved in the normalization of HT are as follows:

1. Determine the absolute values of the IMF and identify all the local maxima of these absolute values
2. Define the envelope by a spline through maxima points. Let this spline be $e_1(t)$.

The normalization is given by

$$f_1(t) = \frac{IMF(t)}{e_1(t)} \tag{2.51}$$

3. To ensure that the value of $f_j(t)$ is strictly lower than unity, the process is to be repeated iteratively (Huang and Wu 2008) as follows

$$f_2(t) = \frac{f_1(t)}{e_2(t)}; \ldots\ldots\ldots f_n(t) = \frac{f_{n-1}(t)}{e_n(t)} \tag{2.52}$$

4. The iteration stops at the nth step, when the normalized maximum values are all unity. Then the empirical FM and AM components of $IMF(t)$ are defined as

$$FM(t) = f_n(t) \tag{2.53}$$

$$AM(t) = \frac{IMF(t)}{FM(t)} \tag{2.54}$$

The steps (1) to (4) yield the empirical AM and FM parts of the IMF. The combination of this normalizing iteration process and application of the Hilbert transform to the empirical AM signal is called the Normalized Hilbert Transform (NHT) method. The use of *arccosine* instead of the *arctan* in the computation of phase angles make the scheme Direct Quadrature (DQ) and the combined procedure is called as NHT-DQ scheme. This scheme is invoked in the study for the analysis of different hydrological time series from India.

The Hilbert transform is a useful mean for characterization of time series signals. Understanding the correlation between two physically linked time series in multiple time scales is of great practical significance. Since most of the hydroclimatic time series possess multiscaling behavior, a scale dependent correlation analysis is more appropriate to establish the link between meteorological variables and hydrological variables (Papadimitriou et al. 2006; Rodo and Rodriguez-Arias 2006; Scafetta 2014). For such analysis wavelet coherence technique was used in the past (Labat 2010; Carey et al. 2013; Araghi et al. 2017), but the selection of mother wavelet may influence wavelet coherence and it may influence subsequent interpretations. Therefore, to capture such associations, using a data adaptive decomposition technique and performing running correlation analysis is more appropriate. But the selection of appropriate window size is a challenging problem while applying running correlation analysis techniques. Stemming from the principles of HHT, Chen et al. (2010) proposed a method, for determining scale dependent correlation between two time series, namely Time Dependent Intrinsic Correlation (TDIC). The algorithm of TDIC analysis is presented in the next section.

2.3.2.7 Time Dependent Intrinsic Correlation Analysis

TDIC method employs the EMD (or its variants) to decompose both of the time series of concern into multiple time scales. It is to be noted that in this method the window size is fixed adaptively (based on properties of data), keeping the stationarity property of the data within the window and more importantly, the size of sliding window is fixed based on the instantaneous period (computed from instantaneous frequencies obtained by HT of IMFs). The different steps involved in TDIC analysis are:

1. Use EMD (or its variants) to decompose the two time series of concern into different time scales
2. Compare the periodicities of IMFs of both series and select those IMFs with nearly same mean periodicity
3. Find the instantaneous periods of both of the IMFs by HT
4. Find the minimum sliding window size (t_d) as maximum instantaneous period between the two signals at the current position t_k, i.e., $t_d = \max(T_{1,i}(t_k), T_{2,i}(t_k))$ where $T_{1,i}$ and $T_{2,i}$ are instantaneous periods
5. Fix the sliding window as $t_w^n = \left[t_k - \dfrac{nt_d}{2} : t_k + \dfrac{nt_d}{2} \right]$

   where n is any positive number (a multiplication factor for minimum sliding window size). In general, n is selected as 1 (Huang and Schmitt 2014).
6. Let IMF 1 and IMF2 are two IMFs of nearly same mean period pertaining to two different time series. The TDIC of the pair of IMFs can be found out as $R_i(t_k, n) = corr(IMF_{1,i}(t_w^n), IMF_{2,i}(t_w^n))$ at any t_k, where $Corr$ is the correlation coefficient of two time series

7. Perform student t-test to investigate whether the difference between the correlation coefficient $R_i(t_k, n)$ and zero is statistically significant or not
8. Repeat steps 4 to 7 iteratively until the boundary of the sliding window exceeds the end points of the time series

After computing the TDIC matrix, the TDIC plot is prepared, where the horizontal axis of the TDIC plot is the time of the series (corresponding to the center position of the sliding window) and the vertical axis is the size of the sliding window. The TDIC plot will be triangular in shape and the correlation at the apex point will be the general correlation coefficient between the series considering the entire data length, if the maximum size of sliding window is fixed as complete data length (Chen et al. 2010). In general, half of the data length is selected as the maximum size of sliding window. The bottom contour of the triangular plots depicts the instantaneous frequency and hence a shift of the plots to larger time scales can be noticed in higher order IMFs (i.e., of low frequency modes). TDIC method is gaining popularity in performing multiscale correlation analysis between teleconnected time series from different fields (Huang and Schmitt 2014, Ismail et al. 2015; Derot et al. 2016). It was also reported that an extended variant of TDIC namely Time Dependent Intrinsic Cross Correlation (TDIC) can also be used for the teleconnection studies with multiple time scales and lags (Chen et al. 2010). In this research, HHT and TDIC methods are employed for understanding the teleconnection between the hydroclimatic time series.

2.4 MEMD-BASED HYBRID FRAMEWORKS FOR HYDROLOGICAL APPLICATIONS

2.4.1 MEMD-TDIC COUPLED FRAMEWORK FOR INVESTIGATING MULTISCALE TELECONNECTIONS

In the past, to investigate the hydroclimatic teleconnections, few studies performed multiscale decomposition of the time-series pair and compared the periodicity of their oscillatory modes, and other studies performed simple statistical correlation analysis (Iyengar and Raghu Kanth 2005; Maity and Nagesh Kumar 2008). The potential of TDIC algorithm enables us to investigate the mutual association between two nonstationarity signals in multiple time scales. This HHT based approach performs a running correlation of oscillatory modes of the time series pair using TDIC analysis. The oscillatory modes are obtained by EMD or its variants. This approach was applied for few marine climatic database and turbulence studies in the past (Huang and Schmitt 2014; Ismail et al. 2015; Derot et al. 2018). However, as multiple climate signals may affect the hydrological processes, and the decomposition of each signal may result in different number of modes, the EMD based scale separation may results in mode-misalignment issues in teleconnection studies, a multivariate approach may be more promising for such studies. In this context, this book presents MEMD-TDIC coupled approach for multiscale teleconnection studies. The different steps involved in the procedure are:

1. Decompose the time series of a hydrological variable (streamflow) and climate indices using the MEMD. With MEMD simultaneous decomposition of multiple signals of concern results in a set of rotational modes of specific periodic properties and distinct characteristics
2. Perform a correlation analysis by finding the correlation coefficient between the IMFs of hydrologic time series with that of a specific climatic index time-series to draw proper inferences regarding the association, in terms of periodicity
3. Rescale the residue of hydrologic time-series about its mean
4. Compare the zero-mean residue of climatic index time series with the zero-mean residue of hydrological time series
5. If the zero crossings of both series are nearly at the same time instant, vital conclusions can be drawn regarding the association of the two signals in terms of nonstationarity
6. Perform TDIC analysis between the respective IMFs (of comparable periodicity) to draw useful inferences on the association between the two in different time scales

A flowchart of the proposed methodology is given in Figure 2.13.

2.4.2 MEMD-BASED HYBRID SCHEME FOR HYDROLOGIC MODELING

Numerous hybrid decomposition models involving DWT/ EMD were proposed in the past for the prediction of hydrologic variables and such models followed distinctly different strategy/scheme while applying for practical datasets (Zhang et al. 2015). Some of the studies prepared individual models for components at different time scales, and finally recombined to get the desired output (Wang et al. 2013, 2015, Huang et al. 2014). In this strategy, they have considered appropriate lags at different timescales for selection of appropriate inputs. Most of the studies followed a hindcasting scheme where the complete dataset chosen is first decomposed, and then separated into calibration and validation datasets. Another scheme, called the real forecasting scheme, is one in which the data division is performed first and then using the calibration data the model is built, subsequently validation data is supplied in blocks, decomposed again to get the predictions. It is reported that the latter method display poor predictive capabilities and suffer from the problems like end effect or boarder distortions, eventhough it is in line with the practical case (Napolitano et al. 2011; Karthikeyan and Nagesh Kumar 2013; Zhang et al. 2015). Only very few studies accounted multiple inputs in the past studies (for example, Wang et al. 2013; Kisi et al. 2014; Zhu et al. 2016) but they used the obtained components directly as input, which is not an effective practice to incorporate the significant inputs at different timescales during the modeling exercise. This book presents an effective framework for prediction of hydrological variables, based on MEMD and SLR, and tested its efficacy of hydrologic predictions following a hindcasting scheme of implementation. The details of the methodology are given below.

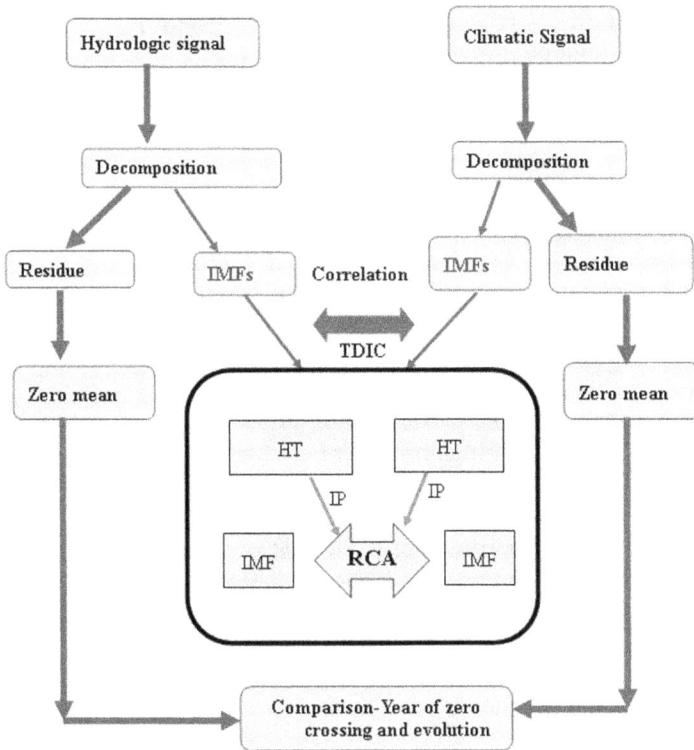

FIGURE 2.13 Flowchart of the methodology for hydroclimatic teleconnection analysis using TDIC approach. RCA refers to running correlation analysis; IP refers to instantaneous period. (Modified from Adarsh and Janga Reddy 2020).

In this framework, MEMD is used for the decomposition of the datasets and uses the stepwise linear regression (SLR) for building the regression models for each of the components. The proposed methodology involves the following steps:

1. Decompose the multivariate data set comprising the output (dependent) variable and different inputs (independent variable) using MEMD to get different orthogonal modes (subseries), called intrinsic mode functions (IMFs), each one is associated with specific time scale of variability.
2. Build SLR models to predict each mode as a function of the corresponding mode of different input variables
3. Refine the models at different time scales after identifying the insignificant components (based on p-value statistic) and discarding such components
4. Predict the modes of output variable at different time scales by using the refined models
5. Recombine the predicted modes to get the output at observation scale

The above built model is designated as MEMD-SLR model. The MEMD-SLR model takes the general form of:

$$RM_{oi} = \sum_{i=1}^{NP} r_i RM_{IVi} \tag{2.55}$$

and the output can be obtained as

$$O = \sum_{i=1}^{M} RM_{Oi} \tag{2.56}$$

where RM denotes a rotational mode (an IMF or residue), M is the total number of decomposed modes; NP is the number of predictor variables; r_i is the regression coefficient; O is the output variable. A flowchart of the proposed methodology for model calibration is provided in Figure 2.14.

It may be noted that for prediction of the different orthogonal modes any of the linear/ nonlinear regression method can be used. The association between orthogonal modes of output variable and input variables is assumed to be linear following

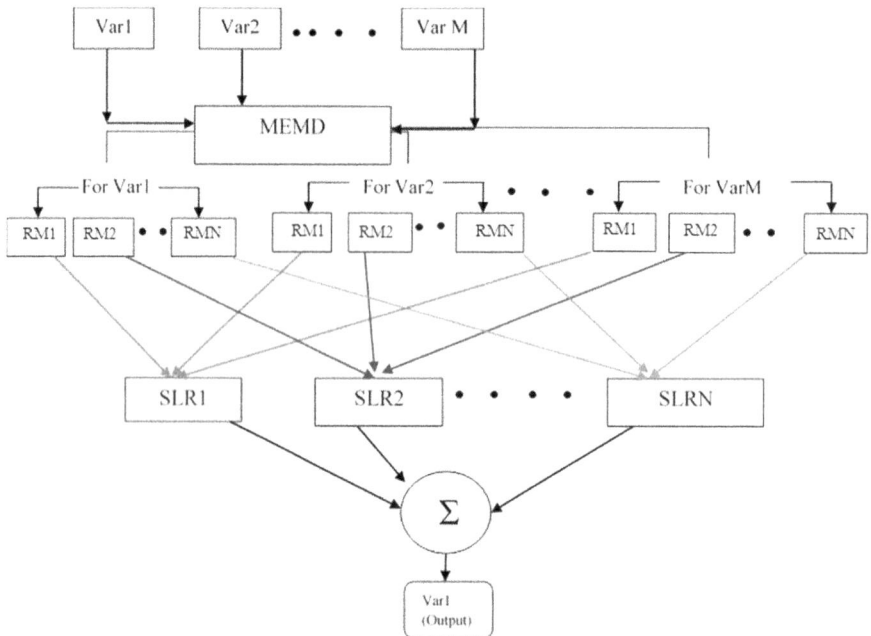

FIGURE 2.14 Flowchart of MEMD-SLR framework for hydrological prediction. (Modified from Adarsh 2017)

Note: Var – Variable; RM-Rotatory Mode; SLR – Stepwise Linear Regression; MEMD – Multivariate Empirical Mode Decomposition.

past studies (Napolitano et al. 2011; Karthikeyan and Nagesh Kumar 2013; Wang et al. 2015) and herein the stepwise linear regression procedure is followed instead of simple linear regression with the impression that the SLR may facilitate the possible exclusion of less significant predictors at different time scales.

The proposed MEMD-SLR framework is general and can be applied for prediction of any hydrological variable. In the next section, its applicability for AISMR prediction is demonstrated.

2.4.3 MEMD-Scaling Theory Coupled Approach for Developing Rainfall Intensity-Duration-Frequency (IDF) Curves

For deriving the IDF relationships based on the scaling property, first the MEMD method is used for the decomposition of rainfall intensity time series of different durations simultaneously. The logarithmic plot between Probability Weighted Moments (PWM) of the orthogonal modes obtained from MEMD and the duration gives the scaling exponent, and finally the IDF relationships are derived based on Extreme Value (EV) formulations involving scaling exponents. Hence the proposed method can be designated as MEMD-EV-PWM method.

In this study, the derivation of IDF curves for subdaily durations is performed using the scale invariance theory. Scaling implies that the statistical properties of the process observed at various scales are governed by the same relationship (Olsson 1998; Olsson and Berndtsson 1998). Let $X(t)$ and $X(\lambda t)$ denotes the observations (time series) at two distinct time scales t and λt following scaling property, it can be represented as

$$X(t) \underline{\underline{dist}} \ \lambda^{-\beta} X(\lambda t) \tag{2.57}$$

where $\underline{\underline{dist}}$ denote the equality of the probability distribution of the two random variables $X(t)$ and $X(\lambda t)$, β is the scaling exponent and λ is the scale conversion factor. According to the scaling theory, a random variable I_d obeys scaling properties if it follows the relation

$$I_d = \left(\frac{D}{d}\right)^{-\beta} I_D \tag{2.58}$$

where D is the aggregated time duration, which can be related to the duration d by defining the scale conversion factor $\lambda_d = \dfrac{D}{d}$, which can be selected as 1, 24 and 720, respectively if we use the hourly data, daily data and monthly data in an exercise of deriving IDF curves for subdaily durations. If the equality in the probability distribution of the random variables (like rainfall depth) observed at two different time scales holds, it implies that the raw moments of any order are also scale invariant. Therefore, if two rainfall intensity time series of duration d and $D = \lambda_d d$ (I_d and I_D, respectively) have similar scaling property, their q order Probability Weighted Moments (PWMs) will bear the following relationship (Kuo et al. 2013):

$$M_d{}^q = (\lambda_d)^{-\beta} M_D{}^q \qquad (2.59)$$

The scaling exponent β can be computed from the slope of log-linear relationship between the PWM of rainfall intensity time series and duration for different moment orders. In addition, in simple scaling, the exponent will be independent of moment order q.

The IDF curves for different return periods can be developed by adopting the following steps:

1. Prepare annual rainfall intensity series for different durations (i.e., 1 day, 2 days, etc.)
2. Decompose all of the rainfall intensity series simultaneously using the MEMD method
3. For each mode, find the PWM values for different durations, for different moment orders
4. Fit a log linear relationship between the PWM and the duration and estimate the slope (β), for each mode, for different moment orders
5. Discard such modes which give low value of coefficient of determination (R^2) statistic (say below 0.8) in the fit or which do not show simple scaling (considerably different β for different moment orders)
6. Form a new series by the addition of all of the remaining modes and repeat steps 2 to 5 until all of the modes pass the criteria mentioned in step 5
7. Compute the average value of the scaling exponent β as the average of slopes for different modes and for different moment orders
8. Develop the IDF curves for different return periods using the extreme value based expression given below (Adarsh 2017):

$$i_d = \frac{\mu + \sigma\left(-\ln\left(-\ln\left(1 - \frac{1}{T}\right)\right)\right)}{d^{-\beta}} \qquad (2.60)$$

where μ and σ are the location parameter and scale parameter respectively given by $\mu = \mu_D(D)^{-\beta}$ and $\sigma = \sigma_D(D)^{-\beta}$; i_d is the rainfall intensity for duration d, T is the return period of i_d. The overall procedure is given in the form of a flowchart, in Figure 2.15.

2.5   CLOSURE

This chapter first presented a brief theoretical background of Fourier transform and Short Time Fourier Transform. Then we looked at a detailed theoretical description on advanced spectral analysis methods such as WT and HHT. The EMD and its variants, HHT and its variants in detail along with the theory of arbitrary order extension of Hilbert spectral analysis, etc. are presented thereafter. The HHT based TDIC is a multiscale running correlation analysis procedure, which can determine

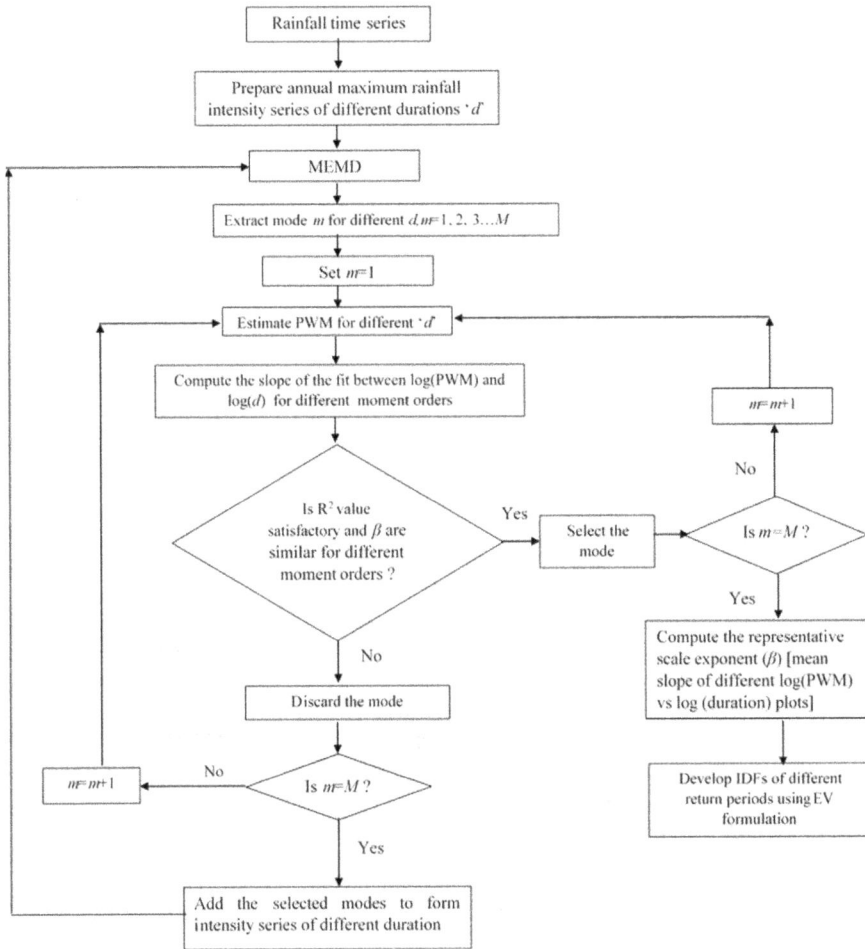

FIGURE 2.15 Flowchart of MEMD-EV-PWM coupled approach for derivation of IDF curves. (From Adarsh 2017).

correlations considering the stationarity within the windows considered along the time domain. This facilitates the application of TDIC to hydroclimatic teleconnection studies. The multi-variate extension of EMD can be considered as a potential tool for different types of practical hydrological applications. The MEMD-TDIC hybrid framework helps for multiscale teleconnection studies in hydrology. The MEMD based hybrid modeling schemes are potential tools for simulations or predictions in hydrology. The MEMD can be coupled with scaling theory to develop the hydrologic design tools like intensity-duration-frequency curves. The algorithmic details of the above three novel frameworks are presented in the chapter. The applications of all of the above techniques for different problems in hydrology are presented in the subsequent chapters.

REFERENCES

Adarsh, S. 2017*Multiscale characterization of hydrologic time series using mathematical transforms*. Ph.D. thesis, IITBombay, India.

Adarsh, S, Janga Reddy, M. 2020. Multiscale modeling of monthly streamflows using MEMD-GP coupled approach. *International Journal of River Basin Management* **18**(2): 139–151.

Araghi, A., Mousavi-Bayg, M., Adamowski, J., Martinezc, C. 2017. Association between three prominent climatic teleconnections and precipitation in Iran using wavelet coherence. *International Journal of Climatology* **37**: 2809–2830.

Aussem, A., Campbell, J., Murtagh, F. 1998. Wavelet-based feature extraction and decomposition strategies for financial forecasting. *Journal of Computational Intelligence and Finance* **6**(2): 5–12.

Bedrosian, E. 1963. A product theorem for Hilbert transforms. *Proceedings of IEEE* **51**(1963): 868–869.

Burrus, C., Gopinath, R., Guo, H. 1998. *Introduction to wavelets and wavelet transforms*. Prentice Hall: New Jersey.

Carey, S.K., Tetzlaff, D., Buttle, J., Laudon, H., McDonnell, J., McGuire, K., Seibert, J., Soulsby, C., Shanley, J. 2013. Use of color maps and wavelet coherence to discern seasonal and interannual climate influences on streamflow variability in northern catchments. *Water Resources Research* **49**: 6194–6207.

Chen, X., Wu, Z., Huang, N.E. 2010. The time-dependent intrinsic correlation based on the empirical mode decomposition. *Advances in Adaptive Data Analysis* **2**: 233–265.

Chou C.M. 2007. Applying multi-resolution analysis to differential hydrological grey models with dual series. *Journal of Hydrology* **332**(1–2): 174–186.

Coifman, R.R., Wickerhauser, M.V. 1992. Entropy-based algorithms for best basis selection. *IEEE Transactions on Information Theory* **38**(2): 713–718.

Daubechies, I. 1990. The wavelet transform, time-frequency localization and signal analysis *IEEE Transactions on Information Theory* **36**(5): 961–1005.

Derot, J., Schmitt, F.G., Gentilhomme, V., Morin, P. 2016. Correlation between long term marine temperature time series from eastern and western English channel: Scaling analysis using empirical mode decomposition method. *Comptes Rendus Geosciece* **348**(5): 343–349.

Grinsted A., Moore J.C., Jevrejeva S. 2004. Application of the cross wavelet transform and wavelet coherence to geophysical time series. *Nonlinear Processes in Geophysics* **11**(5): 561–566.

Harr, A. 1910. Zur Theorie der orthogonalen Funktionensysteme. *Mathematische Annalen* **69**(1910): 331–371.

Hu, W., Si, B-C. (2013). Soil water prediction based on its scale-specific control using multivariate empirical mode decomposition. *Geoderma* 193–194: 180–188.

Huang, G., Su, Y., Kareem, A., Liao, H. 2016. Time-frequency analysis of non-stationary process based on multivariate empirical mode decomposition. *Journal of Engineering Mechanics* **142**(1), 10.1061/(ASCE)EM.1943–7889.0000975.

Huang, N.E., Shen, Z., Long, S.R., Wu, M.C., Shih, H.H., Zheng, Q., Yen, N.C., Tung, C.C., Liu, H.H. 1998. The empirical mode decomposition and the Hilbert spectrum for nonlinear and non-stationary time series analysis. *Proceedings of Royal Society London, Series A* **454**: 903–995.

Huang, N.E., Wu, Z. 2008. A review on Hilbert Huang Transform: Method and its applications to geophysical studies. *Reviews of Geophysics* **46**(2), doi:10.1029/2007RG000228.

Huang, N.E, Wu, Z., Long, S.R., Arnold, K.C., Blank, K., Liu, T.W. 2009b. On instantaneous frequency. *Advances in Adaptive Data Analysis* **1**(2): 177–229.

Huang, S., Chang, J., Huang, Q., Chen, Y. 2014. Monthly streamflow prediction using modified EMD-based support vector machine. *Journal of Hydrology* **511**(2014): 764–775.Huang, Y., Schmitt, F.G. 2014. Time dependent intrinsic correlation analysis of temperature and dissolved oxygen time series using empirical mode decomposition. *Journal of Marine Systems* **130**: 90–100.

Huang, Y., Schmitt, F.G, Lu, Z., Liu, Y. 2009a. Analysis of daily river flow fluctuations using empirical mode decomposition and arbitrary order Hilbert spectral analysis. *Journal of Hydrology* **373**: 103–111.

Huang, Y.X. 2009. *Arbitrary order Hilbert spectral analysis: Definition and application to fully developed turbulence and environmental time series*. Ph.D Thesis in Fluid Dynamics. University of Lille, France.

Ismail, D.K.B., Lazure, P., Puillat, I. 2015. Advanced spectral analysis and cross correlation based on empirical mode decomposition: Application to the environmental time series. *Geoscience and Remote Sensing Letters* **12**(9): 1968–1972.

Iyengar, R.N., Raghu Kanth, T.S.G. 2005. Intrinsic mode functions and a strategy for forecasting Indian monsoon rainfall. *Meteorology and Atmospheric Physics* **90**: 17–36.

Kanwal R.P. 1996. *Linear Integral Equations*. Birkhäuser Boston, Boston, USA.

Karthikeyan, L., Nagesh Kumar, D. 2013. Predictability of non-stationary time series using wavelet and EMD based ARMA models. *Journal of Hydrology* **502**: 103–119.

Kisi, O. 2009. Wavelet regression model as an alternative to neural networks for monthly streamflow forecasting. *Hydrological Processes* **23**(25): 3583–3597.

Kisi, O., Latifoğlu, L., Latifoğlu, F. 2014. Investigation of empirical mode decomposition in forecasting of hydrological time series. *Water Resources Management* **28**(12): 4045–4057.

Klionski, D.M., Oreshko, N.I., Geppener, V.V., Vasiljev, A.V. 2008. Applications of Empirical mode decomposition for processing non-stationary signals. *Pattern Recognition and Image Analysis* **18**(3): 390–399.

Kumar, P., Georgiou, E.F. 1997. Wavelet analysis for geophysical applications. *Reviews of Geophysics* **35**(4): 385–412.

Kuo, C., Gan, T., Chan, S. 2013. Regional Intensity-Duration-Frequency curves derived from ensemble empirical mode decomposition and scaling property. *Journal of Hydrologic Engineering* **18**(1): 66–74.

Labat, D. 2005. Recent advances in wavelet analyzes: Part 1. A review of concepts. *Journal of Hydrology* **314**: 275–288.

Labat, D. 2010. Cross wavelet analyses of annual continental freshwater discharge and selected climate indices. *Journal of Hydrology* **385**: 269–278.

Lee, T., Ouarda, T.B.M.J. 2011. Prediction of climate non-stationary oscillation processes with empirical mode decomposition, *Journal of Geophysical Research* **116**, D06107, doi:10.1029/2010JD015142.

Maheswaran, R., Khosa, R. 2012a. Wavelet-Volterra coupled model for monthly stream flow forecasting. *Journal of Hydrology* **450–451**: 320–335.

Maheswaran, R., Khosa, R. 2012b. Comparative study of different wavelets for hydrologic forecasting. *Computers and Geosciences* **46**(2012): 284–295.

Maity, R., Nagesh Kumar, D. 2008. Basin-scale streamflow forecasting using the information of large-scale atmospheric circulation phenomena. *Hydrological Processes* **22**(5): 643–650.

Mallat, S.G. 1989. A theory for multi-resolution signal decomposition: The Wavelet representation. *IEEE Transactions on Pattern Analysis and Machine Intelligence* **11**(7): 674–693.

Misiti, M., Misiti, Y., Oppenheim, G., Poggi, J.M. 2008. *MATLAB user's guide: wavelet toolbox4*. Natick, MA: Math Works Inc.

Morlet, J., Arens, G., Fourgeau, E., Giard, D. 1982. Wave propagation and sampling theory, Part 1: Complex signal land scattering in multilayer media. *Journal of Geophysics* **47**: 203–221.

Napolitano, G., Serinaldi, F., See, L. 2011. Impact of EMD decomposition and random initialization of weights in ANN hindcasting of daily stream flow series: An empirical examination. *Journal of Hydrology* **406**: 199–214.

Nourani, V., Alami, M.T, Aminfar, M.H. 2009. A combined neural-wavelet model for prediction of Ligvanchai watershed precipitation. *Engineering Applications of Artificial Intelligence* **2**(3): 466–472.

Nourani, V., Baghanam, A.H., Adamowski, J., Gebremichael, M. 2013. Using self-organizing maps and wavelet transforms for space–time pre-processing of satellite precipitation and runoff data in neural network based rainfall–runoff modeling. *Journal of Hydrology* **476**: 228–243.

Nourani, V., Kisi, O., Komasi, M. 2011. Two hybrid artificial intelligence approaches for modelling rainfall runoff process. *Journal of Hydrology* **402**(2011): 41–59.

Nuttall, A.H. (1966). On the quadrature approximation to the Hilbert transform of modulated signals, *Proceedings of IEEE* **54**: 1458–1459.

Olsson, J. 1998. Evaluation of a scaling cascade model for temporal rainfall disaggregation. *Hydrology and Earth System Science* **2**: 19–30.

Olsson, J., Berndtsson, R. 1998. 3. *Water Science and Technology* **37**(11): 73–79.

Pandey, G., Lovejoy, S., Schertzer, D. 1998. Multifractal analysis of daily river flows including extremes for basins five to two million square kilometers, one day to 75 years. *Journal of Hydrology* **208**: 62–81.

Papadimitriou, S., Sun, J., Yu, P.S. 2006. Local correlation tracking in time series. *Proceedings of IEEE Sixth International Conference on Data Mining*, 18–22 December, Hong Kong, pp. 456–465.

Rehman, N., Mandic, D.P. 2010a. Empirical mode decomposition for tri-variate signals. *IEEE Transactions on Signal Processing* **58**(3): 1059–1068.

Rehman N., Mandic, D.P. 2010b. Multivariate empirical mode decomposition. *Proceedings of the Royal Society* Series A **466**(2117): 1291–1302.

Rilling, G., Flandrin, P., Goncalves, P., Lilly, J.M. 2007. Bivariate empirical mode decomposition. *Signal Processing Letters* **14**(12): 936–939.

Rodo, X., Rodriguez-Arias, M.A. 2006. A new method to detect transitory signatures and local time/space variability structures in the climate system: the scale-dependent correlation analysis. *Climate Dynamics* **27**: 441–458.

Rudi, J., Pabel, R., Jager, G., Koch, R., Kunoth, A., Bogena, H. 2010. Multiscale analysis of hydrologic time series data using the Hilbert-Huang Transform. *Vadose Zone Journal* **9**: 925–942.

Sang, Y., Singh, V.P., Sun, F., Chen, Y., Liu, Y., Yang, M. 2016. Wavelet-based hydrological time series forecasting. *Journal of Hydrologic Engineering* 10.1061/(ASCE)HE.1943–5584.0001347, 06016001.

Sang, Y.F. 2013. A review on the applications of wavelet transform in hydrology time series analysis. *Atmospheric Research* **122**(2013): 8–15.

Sang, Y.F., Wang, Z., Liu, C. 2013. Discrete wavelet-based trend identification in hydrologic time series. *Hydrological Processes* **27**: 2021–2031.

Scafetta, N. 2014. Multi-scale dynamical analysis (MSDA) of sea level records versus PDO, AMO, and NAO indexes. *Climate Dynamics* **43**: 175–192.

Torrence, C., Compo, G.P. 1998. A Practical guide to wavelet analysis. *Bulletin of American Meteorological society* **79**(1): 61–78.

Torres, M.E., Colominas, M.A., Schlotthauer, G., Flandrin, P. 2011. A complete ensemble empirical mode decomposition with adaptive noise. *IEEE International conference on Acoustic Speech and Signal Processing*, Prague 22–27 May 2011, pp. 4144–4147.

Wang, W., Ding, J. 2003. Wavelet network model and its application to the prediction of hydrology. *Nature and Science* **1**(1): 67–71.

Wang, W.C., Chau, K.W., Xu, D.M., Chen, X.Y. 2015. Improving forecasting accuracy of annual runoff time series using ARIMA based on EEMD decomposition. *Water Resources Management* **29**(8): 2655–2675.

Wang W.C., Xu D.M., Chau K.W., Chen S. 2013. Improved annual rainfall-runoff forecasting using PSO–SVM model based on EEMD. *Journal of Hydroinformatics* **15**(4): 1377–1390.

Wei, S., Song, J., Khan, N.I. 2012. Simulating and predicting river discharge time series using a wavelet-neural network hybrid modelling approach. Hydrological Processes **26**: 281–296.

Wu, Z., Huang, N.E. 2004a. A study on the characteristics of white noise using the empirical mode decomposition method. *Proceedings of the Royal Society A: Mathematical, Physical and Engineering Sciences* **460**: 1597–1611.

Wu, Z., Huang, N.E. 2004b. Statistical significance test of intrinsic mode functions. In *Hilbert-Huang Transform and Its Applications*. Edited by: Norden E. Huang (NASA Goddard Space Flight Center, USA), Samuel S.P. Shen (University of Alberta, Canada). World Scientific Publishing, Singapore. pp. 149–169.

Wu, Z., Huang, N.E. 2005. Ensemble empirical mode decomposition: A noise-assisted data analysis method. *Center for Ocean-Land-Atmosheric Studies* Calverton, Maryland. Tech. Rep. 193, 1–51.

Zhang, X., Peng, Y., Zhang, C., Wang, B. 2015. Are hybrid models integrated with data preprocessing techniques suitable for monthly streamflow forecasting? Some experiment evidences. *Journal of Hydrology* **530**: 137–152.

Zhou, Y, Leung, Y, Ma, J-M. 2013. Empirical study of the scaling behavior of the amplitude-frequency distribution of the Hilbert Huang transform and its application in the sunspot time series analysis. *Physica A: Statistical Mechanics and its Applications* 392: 1336–1346.

Zhu, S., Zhou, J., Ye L., Meng, C. 2016. Streamflow estimation by support vector machine coupled with different methods of time series decomposition in the upper reaches of Yangtze River, China. *Environmental Earth Science* **75**: 531, doi: 10.1007/s12665-016-5337-7.

3 Wavelet Transform Applications for Hydrological Characterization

3.1 BACKGROUND

Advanced spectral analysis methods are useful for diverse practical applications in hydrology. The hydrological time series are characterized by the complex features like trend, periodicity and shift, which constrains the modeller to search for alternative modelling practices for their simulation and prediction. This chapter of the book considers the applications of more popular wavelet transforms in hydrology. Discrete and continuous variants of wavelet transforms have their own domains of applications in hydrology. DWT has more direct applications in the predictive modelling, owing to its capability to decompose the time series into different components with discrete periodic scale. Such applications are well debated in literature and the applications using emerging methods like HHT with multichannel signals being more appealing to scientific research, this chapter deals with the applications of the DWT and CWT for two specific applications on trend detection and periodicity estimation of complex hydrologic signals. The nonparametric methods like Mann-Kendall (MK) method (Mann 1945; Kendall 1975) and Sen's slope estimator (Sen 1968) are the most popular methods for trend analysis of hydrological signals. The CWT has specific potential in capturing the periodicity of the signals and its extended variants such as wavelet coherence and cross-wavelet spectra has practical usefulness in teleconnection studies. The application of DWT for trend analysis is demonstrated with post-monsoon rainfall in the state of Kerala India, whereas the application of CWT for streamflow-suspended link is demonstrated with the monthly mean data of Kallada river in Southern Kerala.

3.2 DWT APPLICATION FOR TREND ANALYSIS OF RAINFALL

3.2.1 STUDY AREA AND RAINFALL CHARACTERISTICS

The Indian Institute of Tropical Meteorology (IITM) (www.tropmet.res.in) Pune defined the boundaries of 36 meteorological subdivisions in India. The meteorological subdivisions of India (shown in Figure 3.1) are having distinctly different

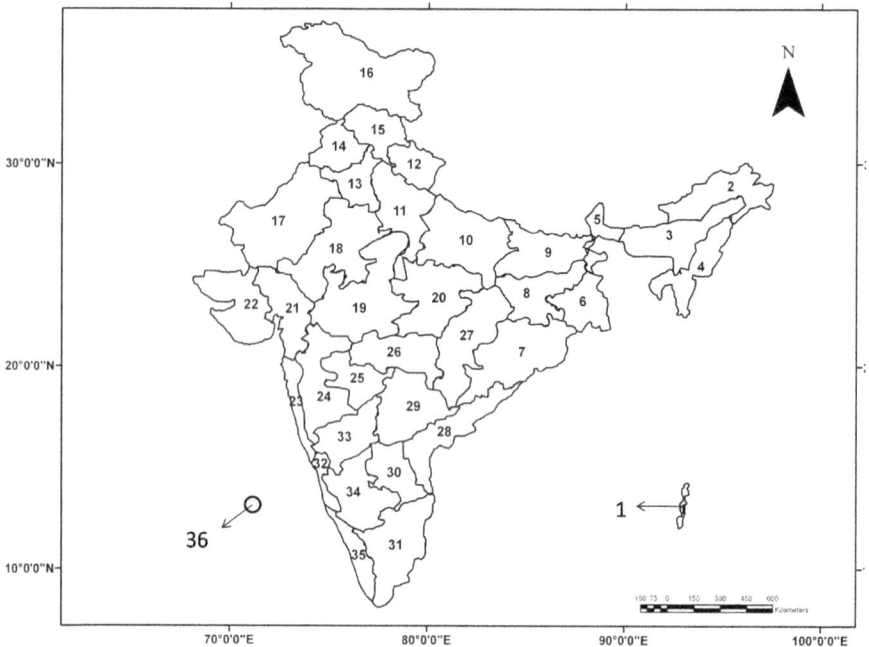

FIGURE 3.1 Location map of meteorological subdivisions in India (1. Andaman Nico bar Islands; 2. Arunachal Pradesh; 3. Assam & Meghalaya; 4. Nagaland, Manipur, Mizoram & Tripura; 5. Sub-Himalaya, West Bengal & Assam; 6. Gangetic West Bengal; 7. Orissa; 8. Jharkhand; 9. Bihar; 10. East Uttar Pradesh; 11. West Uttar Pradesh; 12. Uttaranchal; 13. Haryana, Chandigarh & Delhi; 14. Punjab; 15. Himachal Pradesh; 16. Jammu & Kashmir; 17. West Rajasthan; 18. East Rajasthan; 19. West Madhya Pradesh; 20. East Madhya Pradesh; 21. Gujarat; 22. Saurashtra, Kutch & Diu; 23. Konkan & Goa; 24. Madhya Maharashtra; 25. Marathwada; 26. Vidarbha; 27. Chattisgarh; 28. Coastal Andhra Pradesh; 29. Telangana; 30. Rayalaseema; 31. Tamil Nadu & Pondicherry; 32. Coastal Karnataka; 33. North Interior Karnataka; 34. South Interior Karnataka; 35. Kerala; 36. Lakshadweep).

characteristics with wide variability in rainfall patterns, such as, higher rainfall receiving regions (subdivisions in northeastern parts of India) and relatively lower rainfall receiving regions (subdivisions in northwestern parts of the country).

IITM Pune developed spatially averaged rainfall datasets at subdivisional scale, based on a raingauge network comprising 1416 stations (at least one representative station per district) (Parthasarathy et al. 1995). The state of Kerala ($8°N–12°N$, $74°E–77°E$), popularly known as 'gateway of Indian monsoon', is one of the highest rainfall-receiving regions in India. Kerala is having a coastal belt of more than 700 km and the climate of Kerala is highly influenced by the circulations in Arabian Sea and Indian Ocean. The rainfall trend in Kerala and spectral homogeneity was widely investigated in the past (Krishnakumar et al. 2009; Azad et al. 2010; Adarsh and Janga Reddy 2015). The principal rainy seasons in Kerala are the southwest monsoon (June–September) and the northeast monsoon (October–November). The premonsoon months (March–May) are characterized by major thunderstorm activity in the region, and the winter months (December–February) are marked by low clouding and low rainfall (Ananthakrishnan et al. 1979). The monthly rainfall data of Kerala subdivision for the period 1871–2012 are collected from IITM Pune and seasonal rainfall are computed for each region by adding the monthly rainfall values of the postmonsoon period and used in the present study. The preliminary trend analysis of different seasonal rainfall data of Kerala performed using MK (Mann 1945; Kendall 1975) and Sen's slope estimator (Sen1968) resulted in a Z-statistics of 2.43 and 0.79 at significance level of 5% for the rainfall of post-monsoon season. This shows that the post-monsoon rainfall of Kerala shows statistically significant increasing trend, which was supported in many studies (Krishnakumar et al. 2009; Adarsh and Janga Reddy 2015). To examine the characteristics of the trend of post-monsoon rainfall indepth, the DWT is applied. The Sequential Mann Kendall (SQMK) test (Modarres and Sarhadi 2009) is a technique used for capturing the temporal evaluation of trend, which can be used in conjunction with DWT for the trend analysis (Partal and Küçük 2006; Nalley et al. 2012).

3.2.2 DWT-SQMK Coupled Approach for Trend Analysis

The DWT-SQMK hybrid approach for trend analysis estimation of hydrological time series involves the following steps:

1. Perform the MK and SQMK tests of the time series
2. Select appropriate mother wavelet function and decomposition level
3. Decompose the time series to selected level
4. Prepare different discrete wavelet combination (DWC) series and perform SQMK test of each combination series
5. Perform MK and SQMK test of each DWC series
6. Compare the SQMK values of original time series and DWC series by estimating correlation coefficient (R), Root Mean Square Error (RMSE) and plotting; also compare the MK values of original series and DWC
7. Select few candidate DW combinations for which the SQMK plot is harmonious with that of original series and possess less RMSE and high R value. Identify the periodic scales of the detail (D) components dominantly present in the candidate DW combinations. Identify the most harmonious DW combination with least RMSE, highest R Value and MK value closely match with that of original series

Proper selection of a mother wavelet function is an important step in DWT based trend analysis. Sang et al. (2013) presented 126 types of and the criteria for selection of a wavelet for data preprocessing or characterization of time series include self-similarity, compactness, and smoothness. The Daubechies family is widely adopted in the past for trend analysis studies (deArtigas et al. 2006, Nalley et al. 2012). The Symlet family of wavelets (sym$N$, where N is the number of vanishing moments) is near symmetric and its other properties are similar to those of the most popular Daubechies (db) wavelets. The symlet wavelets are also known as Daubechies' least-asymmetric (LA) wavelets and it has also been adopted in many studies in the past (Kallache et al. 2005, Xu et al. 2009, 2013) for hydrological applications. The symlet wavelets also show characteristics of orthogonality and it provides a compact support indicating that the wavelets have nonzero basis functions over a finite interval, as well as full scaling and translational orthonormality properties (Popivanov and Miller 2002; deArtigas et al. 2006) and these features are very important in localizing events in time-dependent signals (Popivanov and Miller 2002). It has highest number of vanishing moments for a given support width and its associated scaling filters are near linear phase filters (Nibhanupudi 2003).

For the selection of optimum decomposition level of wavelets, a number of criteria were proposed by different researchers, and most of them depend on data length (Aussem et al. 1998; Wang and Ding 2003; Nalley et al. 2012). But Nourani et al. (2011) provided an important recommendation that the selection should also be based on seasonality of the data. For the selection of appropriate decomposition level, Partal and Küçük (2006) selected 5 levels for a series comprising 57 data points; Li et al. (2013) used 7 levels for a dataset comprising 286 data points. In the recent past, some of the researchers performed a rigorous trial and error exercise for selecting appropriate decomposition level, apart from the data length consideration (Nourani et al. 2009a, 2009b, 2011, 2013, 2014, Maheswaran and Khosa 2012a, 2012b). Therefore, a trial and error exercise involving three steps, which simultaneously consider the inherent trend of the data, data length and the characteristics of mother wavelet is adopted to choose the wavelet type and decomposition level. The procedure adopted is presented below:

(i)   the equation used by Kaiser (1994) is used for finding the number of levels of decomposition L:

$$L = \frac{\log\left(\dfrac{n}{2v-1}\right)}{\log(2)} \tag{3.1}$$

where v is the number of vanishing moments of a wavelet and n is the number of data points for the time series which should be selected as the next higher dyadic scale to the length of the dataset. In this study, as the length of the series is 141, n is considered as 256 ($2^8 = 256$). The value of decomposition levels for db2-db10 (and sym2-sym10) are 7, 6, 6, 5, 5, 5, 5, 4 and 4, respectively. Some studies in the past reported that smoother wavelets (that have more vanishing moments) are better at detecting long-term time-varying behavior (good frequency-localization properties) (Adamowski et al. 2009) and many studies in the past used smoother mother wavelets (Kallache et al. 2005 used sym8; de Artigas et al. 2006 used db7).

(ii) the approximation component for the maximum level is determined for a pool of different family of smoother wavelets suggested in the past literature.

(iii) a rigorous trial-and-error exercise is made among different levels 4 to 7 by considering each wavelet type db5-db10 (also for sym5-sym10) to find the

optimum decomposition level. In this exercise, the plot of approximation component is examined to assess the cyclicity (seasonality). If it is cyclic, the next
higher-level approximation is also examined and the level at which the monotonic trend of the data is captured first is selected as the level of decomposition.

The approximation components of different mother wavelets of db family at different
levels are shown in Figure 3.2 and that for symlet family are shown in Figure 3.3,
respectively. This exercise show that while adopting a level 4 or 5, the approximation
component at level four (A4) and five (A5) shows some seasonality (cyclic pattern)
in most of the cases and while adopting A6 level, a monotonically increasing trend
was shown by mother wavelets db7, sym6 to sym8 and sym10. This indicates that the
above wavelets are successful in capturing the increasing trend of postmonsoon rainfall time series of the subdivision and in this study, one of the popular and smoother
sym8 wavelet is adopted for decomposing the time series.

While trying level 7, linear trend of A7 is more visible than that for A6 level. But
it can be inferred that the trend is captured first at the level 6 (where the cyclic pattern
starts vanishing) and it can be selected as the optimum level. This important inference
may be leading to the conclusion that in a modeling exercise using the time series, a
change of level from 6 to 7 may not be leading to perceptible improvement of performance, which is identical to the observation made by Nourani et al. (2011) for
modeling hydrologic time series data. Hence the level at which a clear trend captured
first (i.e., A6) is selected as the optimum decomposition level.

The time series of postmonsoon rainfall was decomposed into various wavelet
subtime series (series of approximation (A) and detail (D)) by discrete wavelet transform. Time series of 2-year mode (D1), 4-year mode (D2), 8-year mode (D3), 16-
year mode (D4), 32-year mode (D5) and 64-year mode (D6) and remainder series
(approximation mode) are obtained. The correlation coefficients between each
subtime series and the postmonsoon rainfall series are computed. The rainfall time
series, time series of different decomposition modes (D1 to D6) and approximation
mode are presented in Figure 3.4. The six levels of decomposition resulted in six variation patterns. The D1 and D2 mode show drastic fluctuations (and hence retains a
large amount of residuals from the raw data). The D3 to D5 modes are smoother than
the initial modes. The 64-year mode (D6) showed that there is one major cycle during
the estimated period. With increase in mode, the hidden increasing trend is more
apparent. The approximation mode (A6) shows that rainfall had a clear increasing
trend for the postmonsoon period.

Furthermore, a total of 63 Discrete Wavelet (DW) combination models are
evaluated to identify the dominancy of the responsible periodic component and all
the DW combinations show an increasing tendency. The SQMK values of different
DW combination models (time series) are computed and the plot of SQMK values
for six basic models (i.e., combining one detail among the series of D1 to D6 with
approximation), are shown in Figure 3.5.

Figure 3.5 shows that the SQMK values of first three series D1+A, D2+A and
D3+A fairly matches with that of original series, as compared to others. It also shows
the dominancy of short-term periodic component (less than a decade) in evaluating the
trend. The Mann-Kendall values are estimated for all the 63 DW combinations series

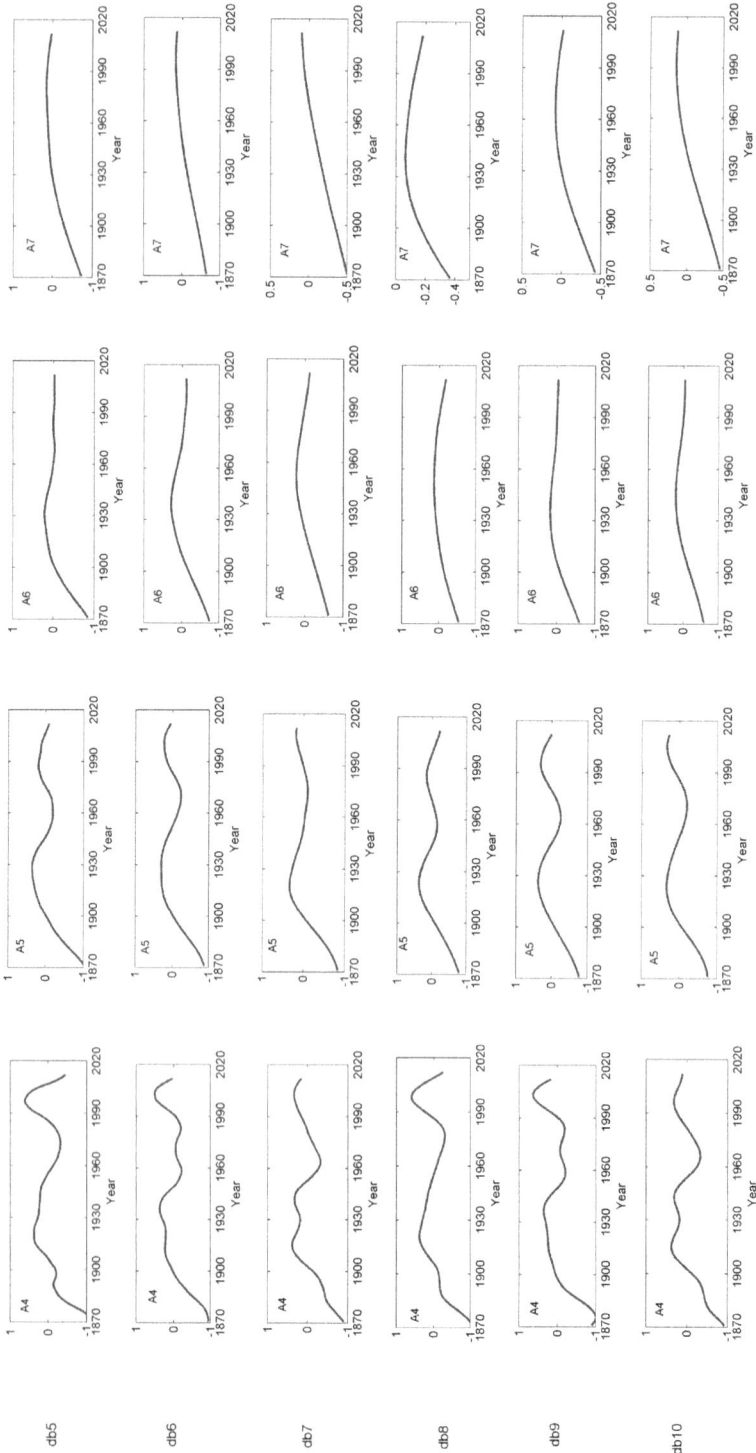

FIGURE 3.2 Approximation components at different levels by different wavelets from Daubechies (dbN) family, for the postmonsoon rainfall of Kerala subdivision.

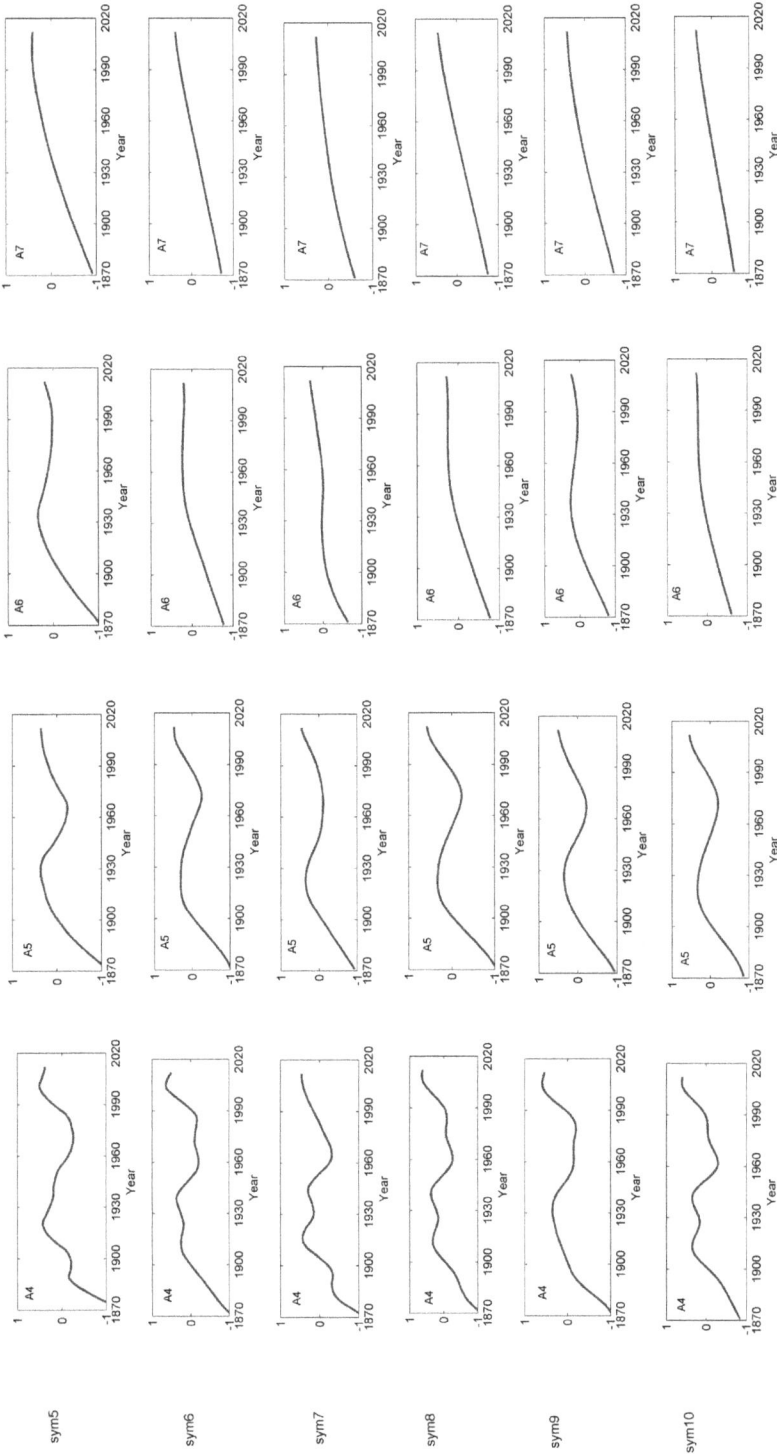

FIGURE 3.3 Approximation components at different levels by different wavelets from symlet (symN) family, for the postmonsoon rainfall of Kerala subdivision.

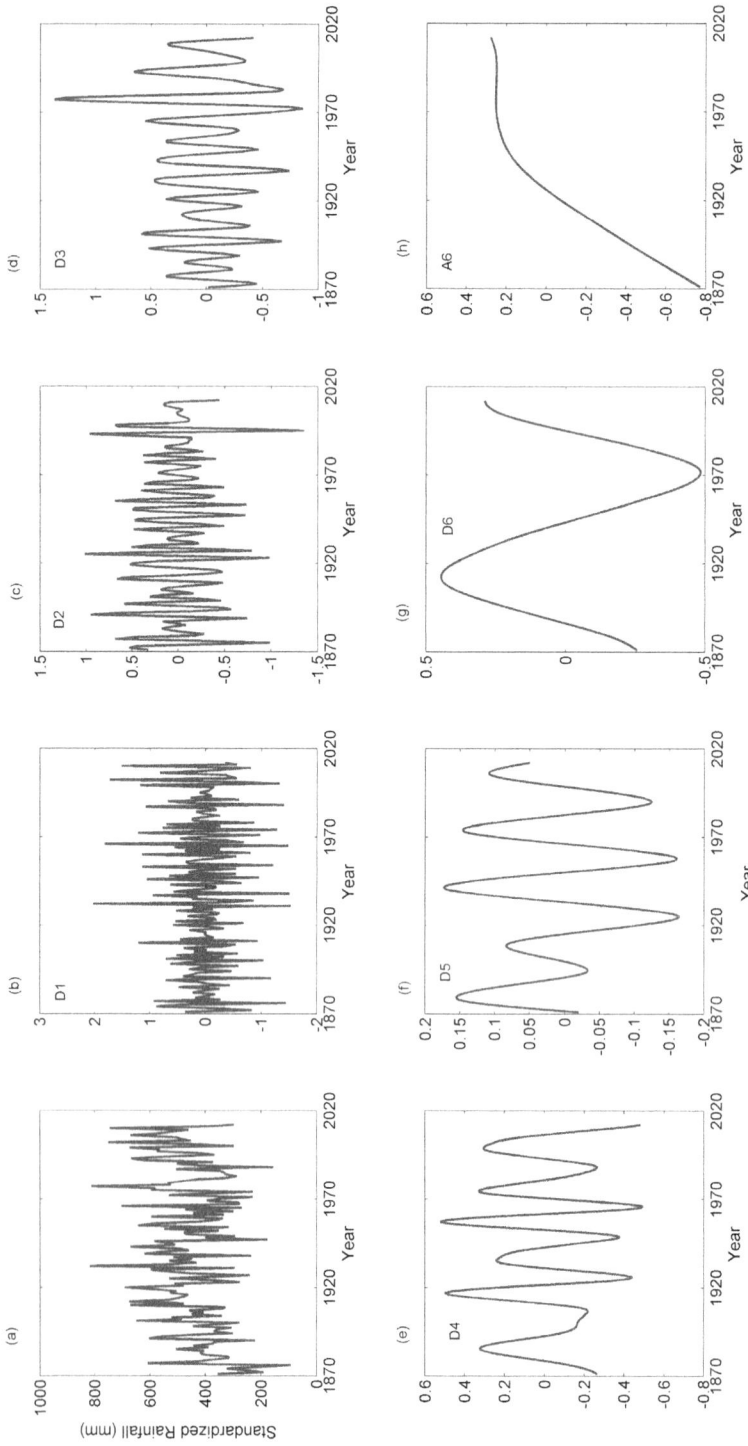

FIGURE 3.4 (a) The time series for postmonsoon rainfall, and (b)–(h) time series for different decomposition modes and approximation mode (A6).

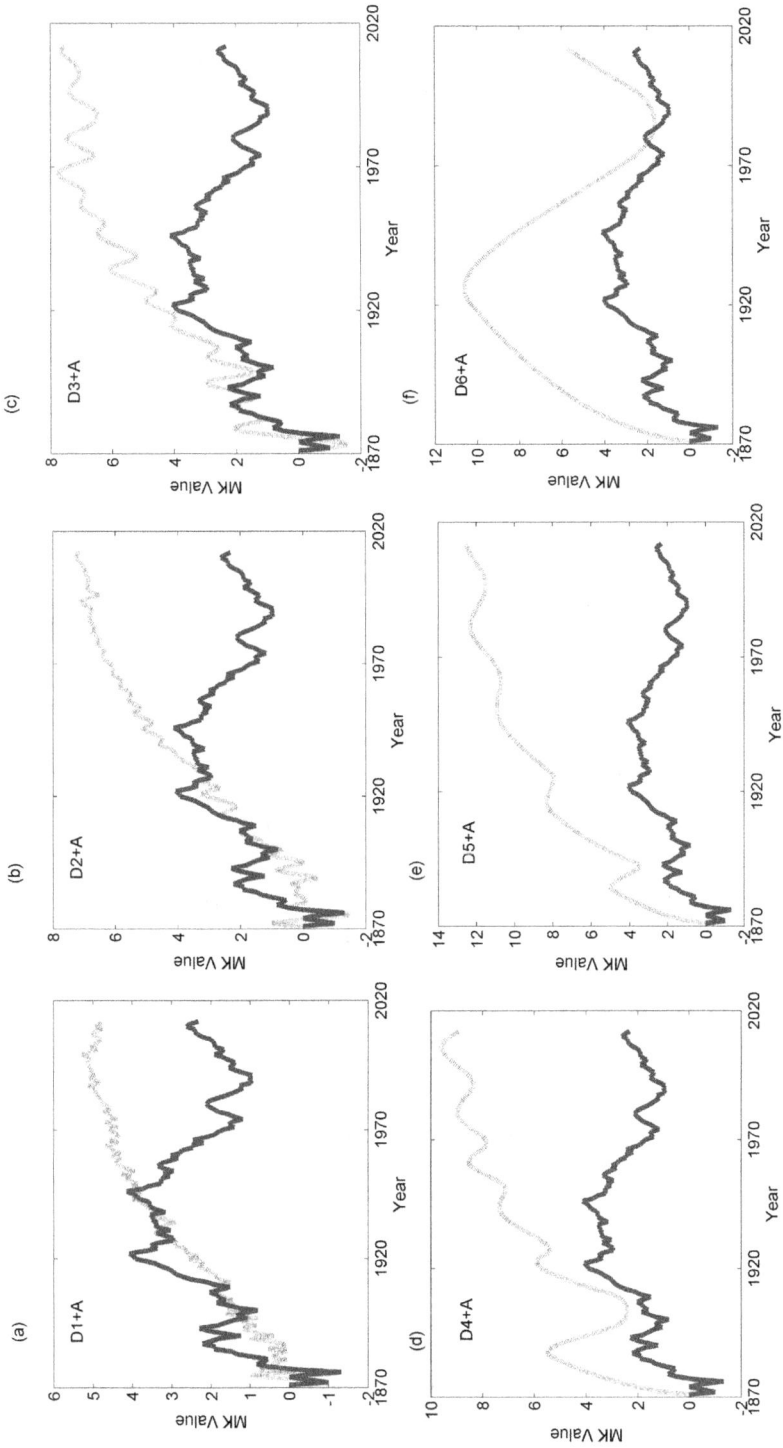

FIGURE 3.5 (a)–(f) Sequential values of MK test for postmonsoon rainfall series (thick line) and basic combination DW models (thin line).

TABLE 3.1
Mann–Kendall (MK) Values of Transformed Series, RMSE and R Between SQMK Values of Original Series and Different Transformed Series

Model	R	RMSE	MK Values
D1+A	0.723	0.692	5.23
D2+A	0.453	0.891	7.49
D3+A	0.464	0.885	8.01
D4+A	0.344	0.938	9.64
D5+A	0.261	0.965	13.08
D6+A	0.379	0.922	5.85
D1+D2+D3+A	0.914	0.406	4.13
D1+D2+D3+A	0.947	0.320	3.91
D1+D2+D3+D5+A	0.921	0.390	4.01
D1+D2+D3+D6+A	0.962	0.271	2.91
D1+D2+D4+D6+A	0.914	0.405	2.85
D1+D3+D4+D6+A	0.910	0.413	3.04
D1+D2+D3+D4+D5+A	0.952	0.307	3.73
D1+D2+D3+D4+D6+A	0.994	0.108	2.67
D1+D2+D4+D5+D6+A	0.920	0.389	2.69
D1+D3+D4+D5+D6+A	0.916	0.399	2.80
D1+D2+D3+D5+D6+A	0.971	0.240	2.76

(transformed series). In addition, the correlation coefficient (R) and root mean square error (RMSE) values between SQMK values of original series and transformed series are computed. The results for 6 basic and 11 mixed DW combinations (only the combinations that show closer matching with original series) are presented in Table 3.1.

Most of the good performing DW combination models contain components D1, D2 and D3 and such combination models have smaller RMSE and higher R values. It is also noticed that the series comprising combinations of the short-term components (D1 to D3) of the subtime series have more influence on the precipitation characteristics of the subdivision. Table 4.6 shows that the MK values of time series formed by D1+D2+D3+D4+D6+A6 model (MK = 2.67) was found to be closest to that of the original series (MK = 2.52). Moreover, this combination has the least RMSE (0.108) and the highest R value as compared to others. The sequential values of the MK test on original series and six constructed models (for models that have reasonably good agreement with the original series) are shown in Figure 3.6.

Figure 3.6 shows closer matching of sequential MK values of D1+D2+D3+D4+D6+A6 model with that obtained for postmonsoon rainfall series. Furthermore, different periodic components are analyzed separately and it is found that only DW3 shows an increasing trend (MK values for D1 to D6 and A6 are −0.4403, −0.0125, 0.0196, −0.073, −1.0035, −3.228, 17.59, respectively). The addition of approximation (A6) component has resulted in increased MK values of D1 to D6 as 0.723,

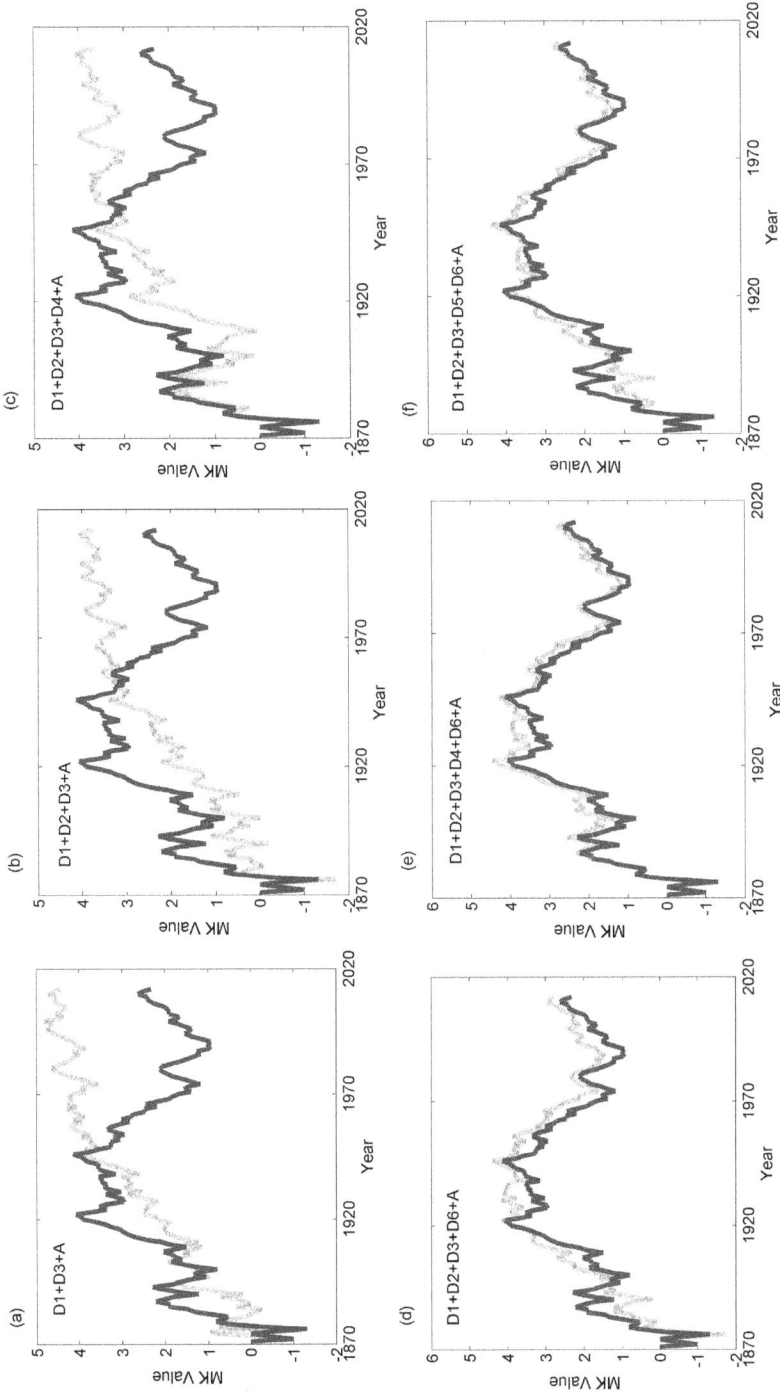

FIGURE 3.6 (a)–(f) Sequential values of the Mann–Kendall (MK) test for postmonsoon rainfall series and constructed models. The thick line represents values for the actual series, and the thin line represents the values for the transformed series.

0.453, 0.464, 0.344, 0.261, 0.379, respectively, even though they have not crossed the significance level value (+1.96). The results show that the approximation component has notable influence on the dataset, as it carries most of the trend component.

In this study, the addition of multiple D components (and approximation) found to enhance the MK value and in some combination the MK value of the reconstructed series is close to that of original series, and the SQMK values of reconstructed series show a harmony with that of original series (also the RMSE value is low). This infers that interaction among the different global phenomena (such as Quasi-Biennial Oscillation (QBO), El Niño Southern Oscillation (ENSO)) having specific periodicity may be influencing the postmonsoon rainfall time series of the subdivision. It is to be noted that rather than the magnitude of the MK value, its nature will also help to arrive at a decision on possible periodicity, subject to the fulfillment of matching of harmony and error criteria (Partal and Küçük 2006; Nalley et al. 2012). Therefore, despite the presence of D1 and D2 in most of the good models (low RMSE, high R value and MK value closer to that of original series), it can be inferred that a periodicity of 8 years is also significant for the postmonsoon rainfall series of Kerala subdivision. Average periodicities of 2–4 years may probably connected with the QBO and 4- to 8-year periodicity can be related to the ENSO phenomenon (Kumar et al. 2013). Earlier researchers proved that the short-term periodicities of 2, 4 and 8 indicated that QBO and ENSO are two prime driving natural processes behind the Indian summer monsoon rainfall (ISMR) (Iyengar and Raghu Kanth 2005; Gadgil et al. 2004; Kumar et al. 2013). The detection of such periodic scales in the postmonsoon rainfall of Kerala meteorological subdivision shows the possible influence of such natural phenomena in rainfall events happening in subsequent months of monsoon period (JJAS).

3.3 APPLICATION OF CWT AND WAVELET COHERENCE FOR ANALYZING STREAMFLOW-SEDIMENT LINK

Multiscale spectral analysis of hydrological time series can be performed efficiently using CWT. Analysis of hydrologic time series using CWT will be more appreciated when such analysis can explain the governing physical processes and possible reasons for variability of the concerned time series. In this perspective, the variability of streamflow and suspended concentration from the Kallada River in Kerala State, India, is analyzed. The monthly time series of streamflow and Total Suspended Sediment (TSS) concentration of Pattazhy gauging station in the Kallada River are first analyzed using the CWT approach and wavelet coherence methods.

3.3.1 STUDY AREA AND DATA DETAILS

Kallada river basin is the largest basin in southern Kerala, India. The major river is Kallada which is formed by the three rivers, i.e., Kulathupuzha, Chendurni and Kalthuruthy joining together near Parappar. The river originates in Papanasam range South of Kulathupuzha in Kollam (Quilon) district of Kerala State at an altitude of 900 m above Mean Sea Level (MSL). The river has a length of 121 km, and drains an area of 1,699 km^2 before joining with Ashtamudi Lake. The Kallada River basin experiences good rainfall and humid atmosphere throughout the year. The climate along the coast of this basin is generally hot with high humidity. The normal daily mean temperature of the basin varies from 26.1°C to 29.1°C. Most of the annual rainfall occurs during June to August season. The annual rainfall varies from 2225 mm to 4038 mm. The Central Water Commission (CWC) maintains one hydrological observation station at Pattazhy on this river. The streamflow and TSS concentration records during 1980–2007 are collected from CWC and used in the present study. The Kallada Irrigation Project (KIP) is the largest irrigation project in the state of Kerala, which originates at the pick-up weir near Ottakkal 5 km downstream of Thenmala Dam (also called Parappar dam by the WRIS India). The dam was commissioned in 1986 and major release started in 1992 and full-fledged functioning began in 2002. The site is selected for the study because a reasonably good length of data prior and after the commissioning of the dam is available from a reliable source. The location map of the Kallada River, the Thenmala dam and the Pattazhi gauging station are shown in Figure 3.7. The photographic view of location of gauging station is provided in Figure 3.8.

3.3.2 WAVELET ANALYSIS OF STREAMFLOW AND TSS CONCENTRATION

The wavelet transform was applied to examine the variation of periodicities in monthly streamflow and sediment load at Pattazhy gauging station. Figure 3.9 presents the Morlet wavelet power spectrum of the monthly streamflow and TSS concentration time series.

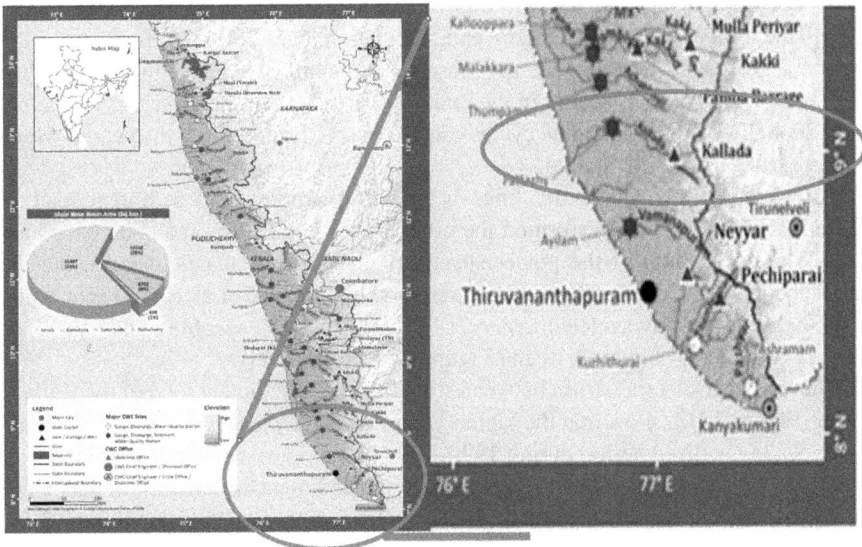

FIGURE 3.7 Location map showing west flowing rivers in southern India, the Thenmala dam and Pattazhy gauging station (Modified from River basin Atlas, WRIS, India).

FIGURE 3.8   The location of Pattazhy gauging station. (From http://india-wris.nrsc.gov.in/
wrpinfo/index.php?title=Pattazhy_HO_46).

It can be clearly seen that both streamflow and sediment load are in good agreement
in both low and high frequency oscillation. The wavelet power spectra indicates intermittent periodicities of 0.5 year and 1 year, which represents the seasonal and annual
alternation of the hydrological variables at Pattazhy station. Furthermore, for streamflow time series, the 95% confidence intervals are continuously distributed at annual
band during 1980–1985 and 1988–2000. Periodicity was not detected for a relatively
short time spell 1986 to 1988 (approximately) and it vanishes completely after 2000.
It can be inferred that the Thenmala dam, completed in the year 1986, affected the
weak periodicity of sediment concentration series, but the periodic nature of streamflow is not affected by the dam commissioning. This could be because that the sediment load might be affected strongly by the human interventions and regulations,
but the natural causes (such as precipitation) might also influencing the periodic
property of streamflow. The complete disappearance of annual cycle from both the
series occurred since 2000 and this can be attributed to the intensive human activities in the upper reaches of Pattazhy gauging station since 2000. In other words, the
human activities have influenced the seasonal and annual cycles of monthly hydrologic series and became weaker in the mid-1990s, and even disappearing completely
after 2000. Also, no intra-annual cycles can be observed since 1992, the year in which
major release to the KIP started.

Furthermore, to analyze the links between streamflow and sediment load of
Kallada river, this study performed the cross wavelet spectrum and coherence analysis. The covariation of the power spectra of the series was calculated using the
wavelet coherence MATLAB package presented by Grinsted et al. (2004) (www.pol.
ac.uk/home/research/waveletcoherence/). The cross wavelet spectrum and coherence
plot are provided in Figure 3.10 and Figure 3.11, respectively.

The cross wavelet transform between streamflow and sediment load of the Kallada
River (Figure 3.10), show that the higher values of cross wavelet power in the annual
band. Some localized period during 1990–1994 higher values of cross wavelet power
is noticed at some intra-annual scales, where the sediment concentration gets more
when the highest flow is recorded. The wavelet coherence analysis (Figure 3.11)
showed higher wavelet coherence at intra-annual to inter-annual scales, which reveal
a stronger relationship between the two series in the time-frequency domain. Here
the in-phase coherence relations are dominant between sediment load and streamflow

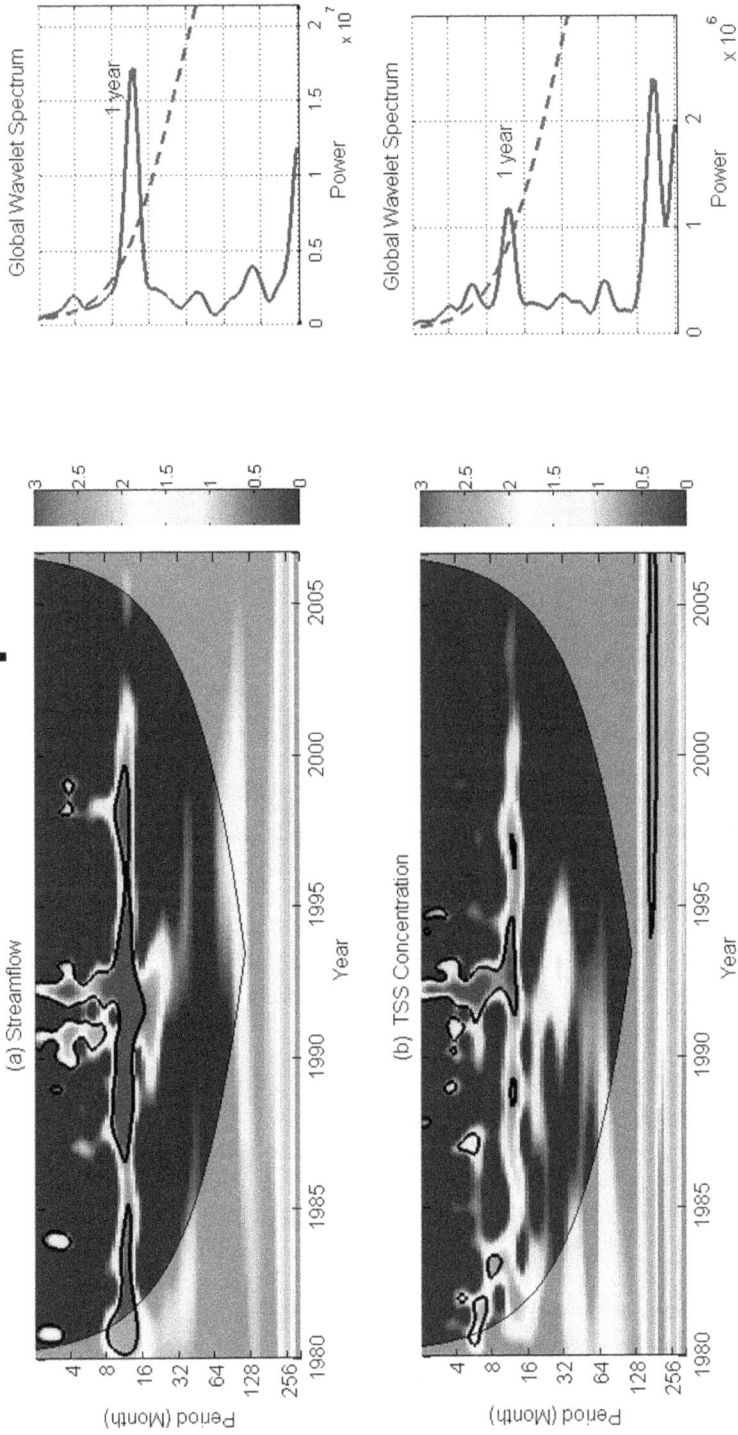

FIGURE 3.9 Wavelet spectrum of (a) Streamflow (b) TSS concentration time series from Pattazhy and corresponding global wavelet power spectra (right panels). The U-shaped line shows cone of influence. The color bar indicates wavelet power and the thick solid line denote 95% confidence level using red noise mode.

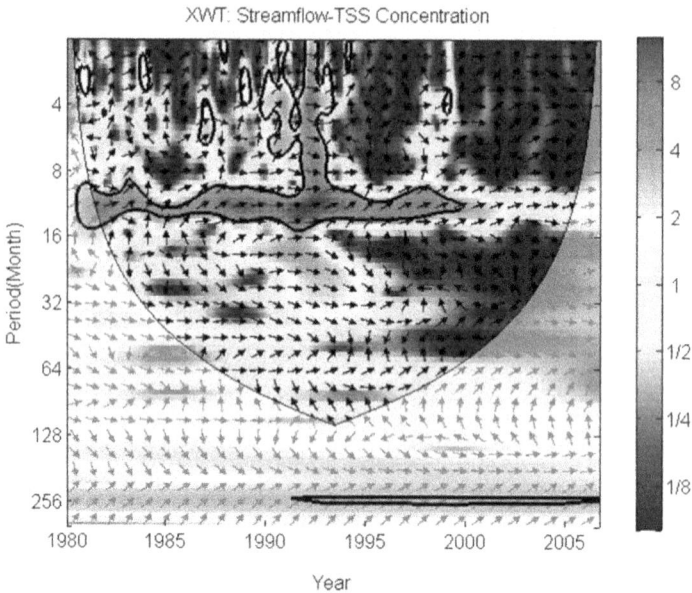

FIGURE 3.10 Cross wavelet spectrum between streamflow and TSS concentration time series of Pattazhy gauging station.

variations at the regions of highest power (indicated from the contour regions like annual band).

Apart from the above analysis, local investigations were made to identify the possible reasons on the variability of the river flow, and it is noticed that: (i) in the past decade, vast sand excavation took place in the Kallada River flowing through Pathanapuram taluk (John 2009), that leads to increase in river depth more than 0.4 m; (ii) prior to 2009 the stream gauging was made by moving boat method and during the period 2009–2010 the float based method was used. Eventhough the former method is more accurate, the excess sand mining offered difficulties for measurement (due to the appearance of rocky strata) and it was forced to switch over to latter method. The bridge based observation of sounding and velocity (using current meter) is followed since 2009–2010; (iii) capacity of Thenmala Dam has been increased, and many distributaries were started operating since 2000 (i.e., a full-fledged operation of the KIP started in 2002) (NRSC 2012); (iv) check dams/temporary weirs were constructed in the upper reaches of Pattazhy; (v) rise in temperature and reduction of rain in the region (John 2009). Hence the effect of human interventions is clearly evident in the hydrological variability of the Kallada River. It is to be noted that WTC analysis (Figure 3.11) showed higher wavelet coherence at intra-annual to inter-annual scales and revealed a stronger positive relationship between the two series in the time-frequency domain. Thus, CWT could successfully capture the influence of anthropogenic interventions on the Kallada River flow.

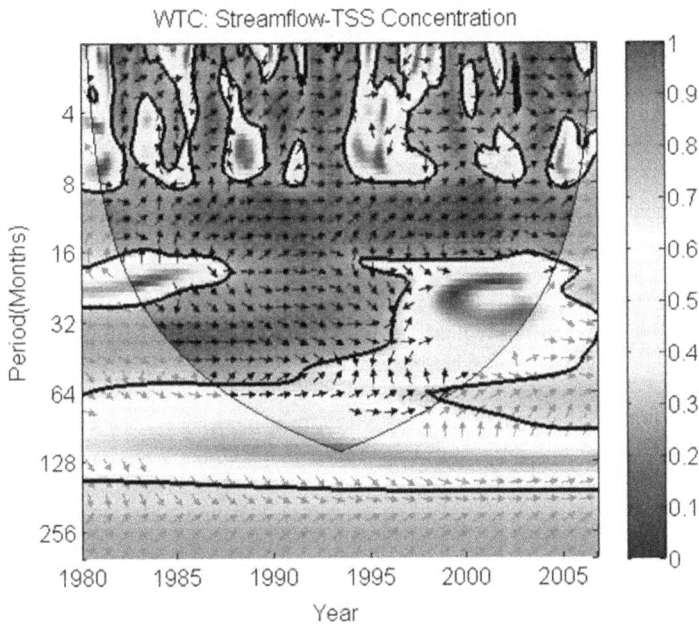

FIGURE 3.11 Squared wavelet coherence between streamflow and TSS concentration time series of Pattazhy gauging station. Arrows indicate the phase difference between the two series of the wavelet spectra (right arrows indicate series are in phase, left arrows indicate series are completely out of phase (180°) and an arrow pointing vertically upward means the second series lags the first by 90° (i.e., the phase angle is 270°). The solid conical curve indicates the cone of influence outside of which paler colors indicate the influence of edge effects and must be viewed with caution. The thicker lines bounding areas of red indicate significant coherence at the 5% level against red noise.

3.4 CLOSURE

This chapter presented two prominent applications of wavelet transforms in hydrology. The DWT in conjunction with SQMK test was found to be a potential tool for detecting the inherent trend in the rainfall time series. This could also capture the dominant periodicity, which could possibly be linked with different large-scale climatic oscillation that can modulate the rainfall processes. The CWT is helpful in detecting the dominant periodic scale, whereas the extensions of CWT has proven to be a potential tool in determining dominant periodicity and capturing the influence of anthropogenic interventions.

REFERENCES

Adamowski, K., Prokoph, A., Adamowski, J. 2009. Development of a new method of wavelet aided trend detection and estimation. *Hydrological Processes* **23**(18): 2686–2696.

Adarsh, S. 2017. *Multiscale characterization of hydrological time series using mathematical transforms*. Ph.D. thesis, IIT Bombay, India.

Adarsh, S., Janga Reddy, M. 2015. Trend analysis of rainfall in four meteorological subdivisions of southern India using non-parametric methods and wavelet analysis. *International Journal of Climatology* **35**(6): 1107–1124.

Ananthakrishnan, R., Parthasarthy, B., Pathan, J.M. 1979. Meteorology of Kerala. *Contributions to Marine Sciences* **60**: 123–125.

Aussem, A., Campbell, J., Murtagh, F. 1998. Wavelet-based feature extraction and decomposition strategies for financial forecasting. *Journal of Computational Intelligence and Finance* **6**(2): 5–12.

Azad, S., Vignesh, T.S., Narasimha, R. 2010. Periodicities in Indian monsoon rainfall over spectrally homogeneous regions. *International Journal of Climatology* **30**: 2289–2298.

de Artigas, M.Z., Elias, A.G., de Campra, P.F. 2006. Discrete wavelet analysis to assess long-term trends in geomagnetic activity. *Physics and Chemistry of Earth* **31**(1–3): 77–80.

Iyengar, R.N., Raghu Kanth, T.S.G. 2005. Intrinsic mode functions and a strategy for forecasting Indian monsoon rainfall. *Meteorology and Atmospheric Physics* **90**: 17–36.

Gadgil, S., Vinayachandran, P.N., Francis, P.A., Gadgil, S. 2004. Extremes of the Indian summer monsoon rainfall, ENSO and equatorial Indian Ocean oscillation. Geophysical Research Letters 31 L12213 doi:10.1029/2004GL019733.

Grinsted, A., Moore, J.C., Jevrejeva, S. 2004. Application of the cross wavelet transform and wavelet coherence to geophysical time series. *Nonlinear Processes in Geophysics* **11**(5): 561–566.

John, E. 2009. The impact of sand mining in Kallada River (Pathanapuram Taluk) Kerala. *Journal of Basic and Applied Biology* **3**(1–2): 108–113.

Kaiser, G. 1994. *A friendly guide to wavelets*. Birkhauser, Boston, MA.

Kallache, M., Rust, H.W., Kropp, J. 2005. Trend assessment: applications for hydrology and climate research. *Non-Linear Processes in Geophysics* **12** (2): 201–210.

Kendall, M.G. 1975. *Rank correlation methods*. Charles Griffin, London.

Krishnakumar, K.N., Rao, G.S.L.H.V.P., Gopakumar, C.S. 2009. Rainfall trends in twentieth century over Kerala, India. *Atmospheric Environment* **43**: 1940–1944.

Kumar, K.N., Rajeevan, M., Pai, D.S., Srivastava, A.K., Preethi, B. 2013. On the observed variability of monsoon droughts over India. *Weather and Climate Extremes* **1**: 42–50.

Li, M., Jun, X., Zhi, C., Meng, D., Xu, C. 2013. Variation analysis of precipitation during past 286 years in Beijing area, China, using non-parametric test and wavelet analysis. *Hydrological Processes* **27**: 2934–2943.

Maheswaran, R., Khosa, R. 2012a. Wavelet-Volterra coupled model for monthly stream flow forecasting. *Journal of Hydrology* **450–451**: 320–335.

Maheswaran, R., Khosa, R. 2012b. Comparative study of different wavelets for hydrologic forecasting. *Computers and Geosciences* **46**(2012): 284–295.

Maheswaran, R., Khosa, R., Adamowski, J., Sudheer, Ch., Partheepan, C., Anand, J., Narsimlu, B. 2014. Wavelet-based multiscale performance analysis: An approach to assess and improve hydrological models. *Water Resources Research* **50**(12): 9721–9737.

Mann, H.B. 1945. Non-parametric tests against trend. *Econometrica* **13**: 245–259.

Modarres, R., Sarhadi, A. 2009. Rainfall trends analysis of Iran in the last half of the twentieth century. *Journal of Geophysical Research* **114**: D03101, doi:10.1029/2008JD010707.

Nalley, D., Adamowski, J., Khalil, B. 2012. Using discrete wavelet transforms to analyze trends in streamflow and precipitation in Qubec and Ontario (1954–2008). *Journal of Hydrology* **475**: 204–228.

Nibhaanupudi, S. 2003. *Signal denoising using wavelets*. M.S. thesis, University of Cincinnati, Ohio.

Nourani, V., Baghanam, A.H., Adamowski, J., Gebremichael, M. 2013. Using self-organizing maps and wavelet transforms for space–time pre-processing of satellite precipitation and runoff data in neural network based rainfall–runoff modeling. *Journal of Hydrology* **476**: 228–243.

Nourani, V., Baghanam, A.H., Adamowski, J., Kisi, O. 2014. Applications of hybrid wavelet artificial intelligence models in hydrology: A review. *Journal of Hydrology* **514**(6): 358–377.

Nourani V., Alami M.T, Aminfar M.H. 2009a. A combined neural-wavelet model for prediction of Ligvanchai watershed precipitation. *Engineering Applications of Artificial Intelligence* 22(3): 466–472.

Nourani, V., Komasi, M., Mano, A. 2009b. A Multivariate ANN-Wavelet approach for rainfall–runoff modeling. *Water Resources Management* **23**: 2877–2894.

Nourani, V., Kisi, O., Komasi, M. 2011. Two hybrid artificial intelligence approaches for modelling rainfall runoff process. *Journal of Hydrology* **402**(2011): 41–59.NRSC. 2012. *Assessment of Irrigation Potential Created in AIBP funded 50 Irrigation Projects in India using Cartosat Satellite Data (Phase-II) –A Report on Kallada Irrigation Project Submitted to Central Water Commission India.* NRSC-RSAA-WRG-AIBP50-Sept 2012-TR-441.

Partal, T., Küçük, M. 2006. Long-term trend analysis using discrete wavelet components of annual precipitations measurements in Marmara region (Turkey). *Physics and Chemistry of the Earth* **31**: 1189–1200.

Parthasarathy, B., Munot, A.A., Kothawale, D.R. 1995. *Monthly and seasonal rainfall series for all-India homogeneous regions and meteorological subdivisions: 1871–1994,* Research Report RR-065, Pune: Indian Institute of Tropical Meteorology.

Popivanov, I., Miller, R.J. 2002. Similarity search over time-series data using wavelets. In: *Proceedings of 18thInternational Conference on Data Engineering*, pp. 212–221.

Sang, Y.F., Wang, Z., Liu, C. 2013. Discrete wavelet-based trend identification in hydrologic time series. *Hydrological Processes* 27: 2021–2031.

Sen, P.K. 1968. Estimates of the regression co-efficient based on Kendall's tau. *Journal of the American Statistical Association* **63**: 1379–1389.

Wang, W., Ding, J. 2003. Wavelet network model and its application to the prediction of hydrology. *Nature and Science* **1**(1): 67–67.

Xu, J., Chen, Y., Li, W., Ji, M., Dong, S., Hong, Y. 2009. Wavelet analysis and non-parametric test for climate change in Tarim river basin of Xinjiang during 1959–2006. *Chinese Geographical Science* **19**(4): 307–313.

Xu, J., Xu, Y., Song, C. 2013. An integrative approach to understand the climatic hydrological processes: A case study of Yarkand River Northwest china. *Advances in Meteorology* ID272715. doi:org/10.1155/2013/2727215.

4 Hilbert Huang Transform Applications for Characterization of Rainfall

4.1 BACKGROUND

The characteristic analysis of rainfall time series involves the determination of the trend, time-frequency spectral analysis, teleconnection and scaling analysis. The variability of the hydrological time series can be assessed by performing a formal trend analysis, which is relevant nowadays due to climate change and its socio-economic consequences. The spatiotemporal variability of rainfall may give a distinctly different pattern of trend in different meteorological subdivisions of India. The decomposition of rainfall time series into multiple orthogonal components can give new insights on its teleconnections with climatic variables and mathematical transformation of the components obtained are performed to get information on its spectral properties, which may eventually help for improved seasonal predictions of rainfall. As the hydrologic and climate signals possess multiscaling properties, it is appropriate to investigate such teleconnections in a multiscaling framework. Moreover, the scaling characteristics from different rainfall intensity time series can be effectively used in estimating the scaling exponent and it can be subsequently used in developing the fine resolution Intensity-Duration-Frequency (IDF) curves from coarse resolution rainfall datasets.

Hilbert Huang Transform (HHT) is a popular method for performing the characteristic analysis of complex time series in the time-frequency domain. HHT and its extensions are helpful in extracting the nonlinear trend, performing the spectral characterization in a time-frequency domain, teleconnections in multiple time scales and for extracting the scaling properties of rainfall. Apart from the popular applications of HHT for trend and spectral analysis, an HHT based framework namely Time Dependent Intrinsic Correlation (TDIC) is presented to investigate the climatic teleconnections of Indian Summer Monsoon Rainfall (ISMR). The implementation of this framework is demonstrated using both single and multivariate EMD (MEMD) approaches. Further, a novel MEMD based framework for developing hourly IDF curves of eight cities in the state of Kerala, India from the respective daily time series is presented. The analysis could have different impacts in different spatial (subdivisions to macro regions) and temporal (hourly data to annual data). Therefore, this chapter considers data of different spatio-temporal domains for demonstrating diverse applications of HHT.

4.2   STUDY AREA AND DATA

All 36 subdivisions defined by IITM Pune together is called as All India (AI) and they defined six nonoverlapping regions namely Hilly Region (HR), Central North East India (CNEI), North East India (NEI), North West India (NWI), Western Central India (WCI), and Peninsular India (PI) based on similarity in rainfall characteristics. The map of nonoverlapping macro regions in India is given in Figure 4.1.

The average rainfall of all India, different nonoverlapping regions and percentage contribution to various seasonal rainfalls are presented in Table 4.1. From Table 4.1 it is noted that at all the macro regions, monsoon rainfall is the dominant contributor to the Average Annual Rainfall (AAR). Further it is noticed that at all the nonoverlapping regions except NEI and PI, the contribution of monsoon rainfall is more than 80% of total rainfall. At NEI, the second major contributor is pre-monsoon rainfall (~21%) while at PI, the second major contributor is the postmonsoon rainfall (~26%).

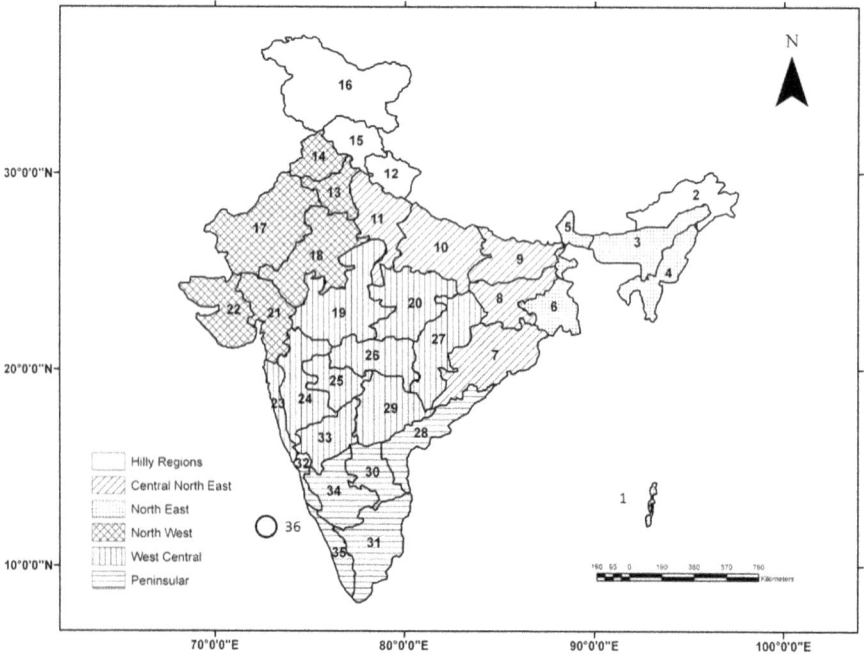

FIGURE 4.1 Map of five nonoverlapping regions in India. The names of different subdivisions against the numbers 1–36 are provided along with Figure 3.1.

TABLE 4.1
Average Annual Rainfall (AAR) and percentage contribution of AAR to different seasonal rainfalls of different macro regions in India

Annual /Season	AI		NEI		CNEI	
	Average Rainfall (mm)	% of AAR	Average Rainfall (mm)	% of AAR	Average Rainfall (mm)	% of AAR
Annual	1086.24	100.00	2056.96	100.00	1197.97	100.00
Monsoon	847.93	78.06	1409.49	68.52	998.69	83.37
Postmonsoon	108.81	10.02	168.30	8.18	84.77	7.08
Winter	35.15	3.24	51.81	2.52	40.62	3.39
Premonsoon	94.35	8.69	427.36	20.78	73.89	6.17
Annual Season	NWI		WCI		PI	
	Average Rainfall (mm)	% of AAR	Average Rainfall (mm)	% of AAR	Average Rainfall (mm)	% of AAR
Annual	545.65	100.00	1070.83	100.00	1163.15	100.00
Monsoon	490.63	89.92	924.33	86.32	659.38	56.69
Postmonsoon	16.24	2.98	77.65	7.25	304.32	26.16
Winter	18.42	3.38	25.97	2.42	61.96	5.33
Premonsoon	20.36	3.73	42.89	4.01	137.49	11.82

4.3 TREND ANALYSIS OF RAINFALL USING NON-PARAMETRIC TESTS AND HHT

The following section presents the results of trend analysis of monsoon rainfall from all India and nonoverlapping regions using different nonparametric methods, EMD and linear fitting. Subsequent section presents the results of trend analysis of annual data, seasonal data and data of different months at subdivisional spatial scale by similar methods.

4.3.1 TREND ANALYSIS OF AISMR AND MONSOON RAINFALL OVER HOMOGENEOUS REGIONS

A number of studies conducted to detect the trends of hydro-climatic variables in different parts of India and reported spatial variability of rainfall is significantly influenced by the physiography of the region (Kumar et al. 2010). The preliminary examination of statistical properties of the spatially average dataset of different macro regions (all India and the five nonoverlapping regions) presented in section 4.2 showed that the major contribution of rainfall is in the summer monsoon season (Table 4.1). Therefore, the monsoon rainfall is considered to get an overall picture of rainfall

TABLE 4.2
Results of MK test and Sen's slope (SS) estimate of monsoon rainfall in six macro regions

Region	MK	Sen's slope
AI	−1.245	−0.223
NEI	−2.460	−0.750
CNEI	−1.303	−0.305
NWI	0.106	0.037
WCI	−1.693	−0.402
PI	0.692	0.095

pattern in India at larger spatial scales. First, the two popular nonparametric methods Modified Mann Kendall (MMK) method (Hamed and Rao 1998) and Sen's slope estimator (Sen 1968) are applied on the all India summer monsoon rainfall (AISMR) series and summer monsoon rainfall of five nonoverlapping rainfall homogenous regions-NEI, CNEI, NWI, WCI and PI and the results are summarized in Table 4.2.

The MK values obtained from the MMK test (at 5% significance level) conducted upon the monsoon rainfall series of six macro regions showed that there is a statistically significant reduction in the monsoon rainfall of NEI, which includes many the highest rainfall receiving locations (such as Chirapunji and Mawsynrum) of Indian subcontinent. At all India level also monsoon rainfall shows a reduction, even though the trend is not statistically significant. The rainfall in the two regions where monsoon contribution is relatively lesser (~69% and ~57%) the trend is found to be increasing even though the trend is not statistically significant. To portray the long term trend, the linear fitting and EMD of the different monsoon rainfall time series are performed and the results are presented in Figure 4.2.

From Figure 4.2 it is clear that a monotonic reduction in the nonlinear trend is noticed in NEI, NWI and WCI, while in the other regions, a quite different pattern of trend is noticed over the study period (1871–2012). It can also be noticed that the inherent nonlinear trend of monsoon rainfall in NWI displays a reduction, while the nonparametric tests detected an increasing trend.

4.3.2 TREND ANALYSIS OF RAINFALL IN KERALA SUBDIVISION

The rainfall pattern in India shows district spatial variability and detailed estimation of trend at smaller spatial resolution like subdivisions/river basins is also important for efficient planning and management of water resources of the country. Also, such analysis is to be carried out at different temporal scales such as annual, seasonal and monthly time scales. First, the linear regression is applied for the trend analysis of rainfall time series of Kerala meteorological subdivision at annual and seasonal scales. For seasonal analysis, each year is divided into four climatic seasons (Ananthakrishnan et al. 1979), namely winter (December–February), pre-monsoon (March–May), southwest monsoon (June–September), and postmonsoon (October–November). The linear trend of annual rainfall and different seasonal rainfalls in Kerala subdivision are presented in Figure 4.3.

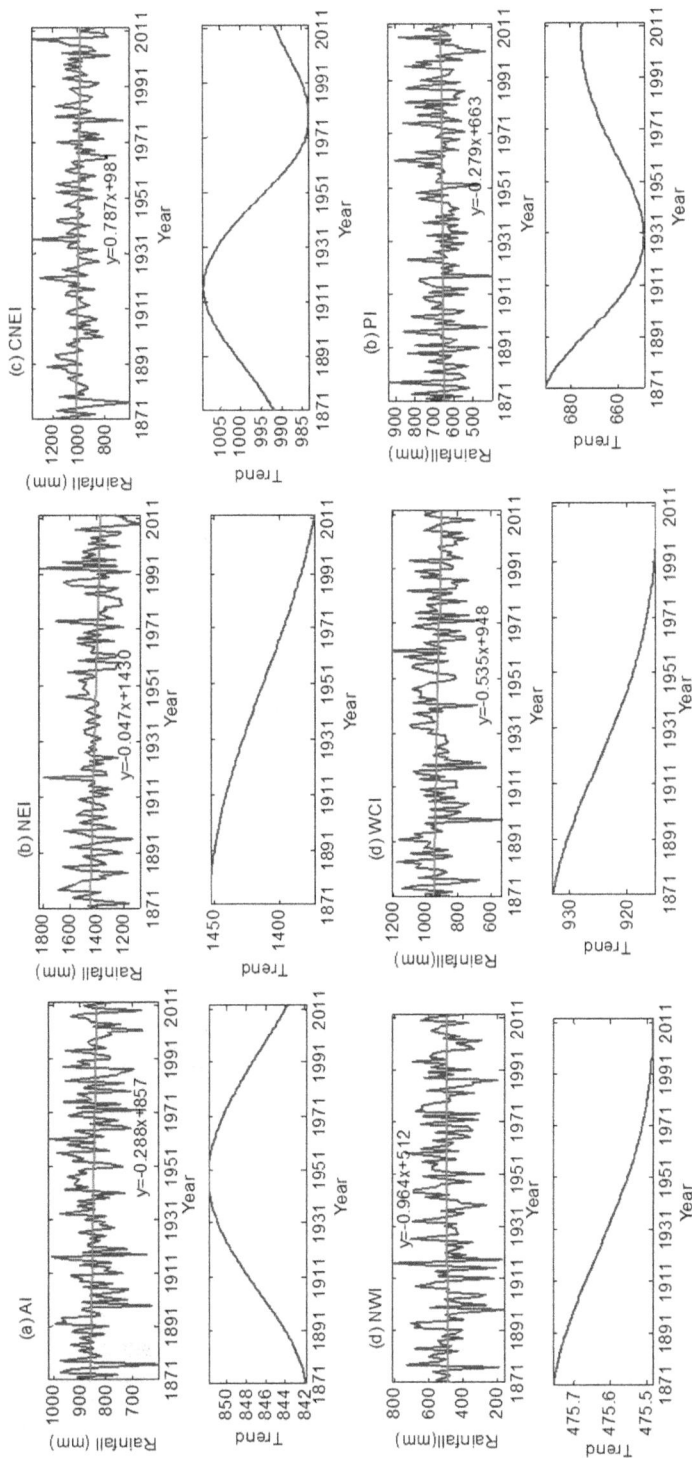

FIGURE 4.2 Linear and nonlinear trend analysis of monsoon rainfall in different macro regions. The lower panels of each case show the nonlinear trend extracted by EMD.

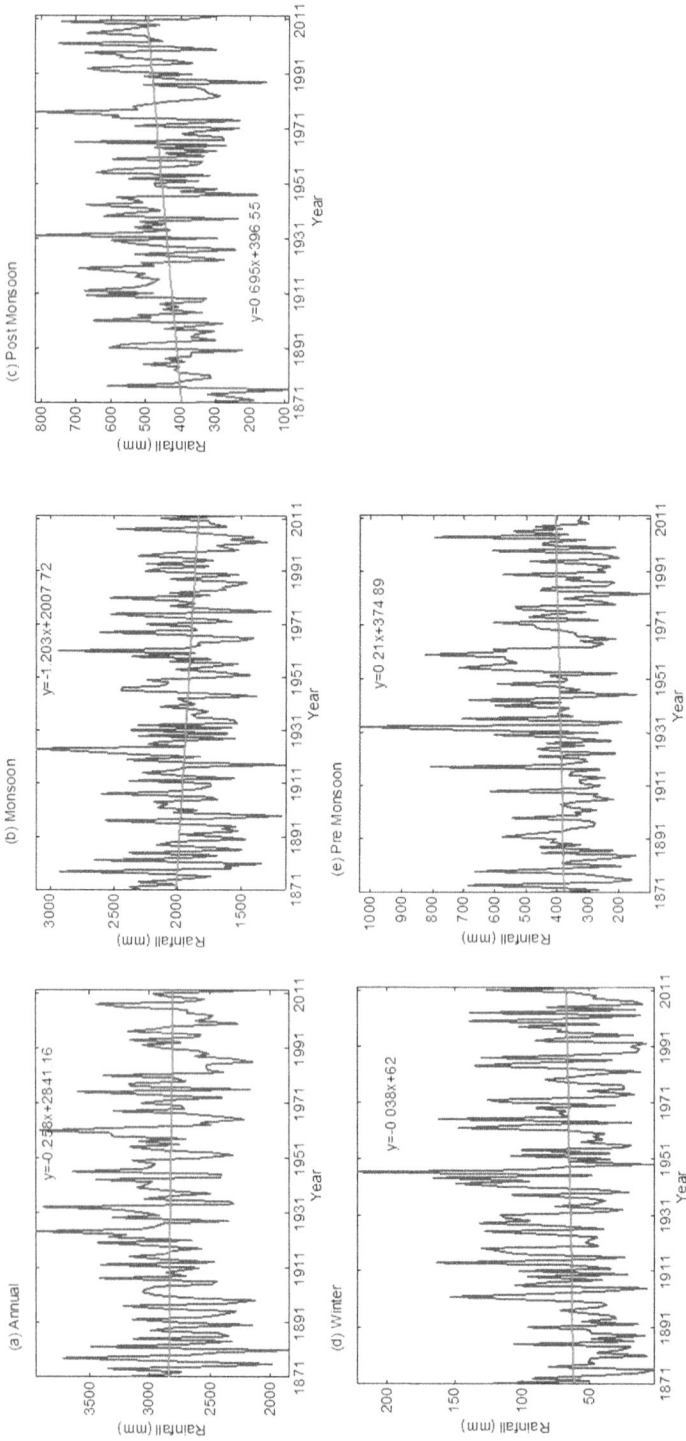

FIGURE 4.3 Linear trend analysis of annual and seasonal rainfalls in Kerala subdivision.

TABLE 4.3
Results of trend analysis of rainfall in Kerala by nonparametric tests

Month/Season	Method	
	SS	MMK
January	0.010	1.659
February	0.027	1.376
March	0.05	0.998
April	0.175	1.611
May	−0.034	−0.128
June	−0.984	**−2.435**
July	−0.584	−1.356
August	0.234	0.925
September	0.412	1.692
October	0.361	1.734
November	0.321	1.905
December	−0.019	−0.451
Monsoon	−1.0797	−1.367
Postmonsoon	0.792	**2.524**
Winter	0.036	0.426
Premonsoon	0.210	0.795
Annual	−0.025	−0.036

Note: Bold numbers show that the trend is statistically significant at 5% significance level.

The results of trend by two-non parametric methods (MK test and SS method) are presented in Table 4.3. Table 4.3 shows that the postmonsoon rainfall of Kerala shows a significantly increasing trend, while the rainfall in June shows a significant reduction, which support earlier findings using a dataset of past century (Krishnakumar et al. 2009).

Despite the difference in data length, few studies in the recent past also reported that there is an increasing trend for post-monsoon rainfall in Kerala subdivision (Guhathakurta and Rajeevan 2008; Krishnakumar et al. 2009; Kumar et al. 2010). But the monsoon and annual rainfall of the subdivision show slightly different response based on data length selected for the study. Hence it is inferred that the contribution in post-monsoon period is less susceptible to changes during the past few decades. In order to identify how the trend gets fluctuated over the study period, the sequential version of Mann Kendall (SQMK) test is useful. The results of SQMK test of the annual and seasonal rainfall of Kerala are provided in Figure 4.4.

Figure 4.4 show that the annual rainfall has multiple intersection points but none of them falls on threshold lines (±1.96 lines), and hence these are not considered as significant turning points. The annual rainfall series show an increasing trend from 1910–2000 and during 2001–2010 the trend is found to be decreasing. However, the increasing trend was significant only for a few short-term spells during the period 1930–1950 and remains significant for a decade from the year 1957. From the SQMK analysis, it is inferred that (i) rainfall time series of Kerala showed statistically

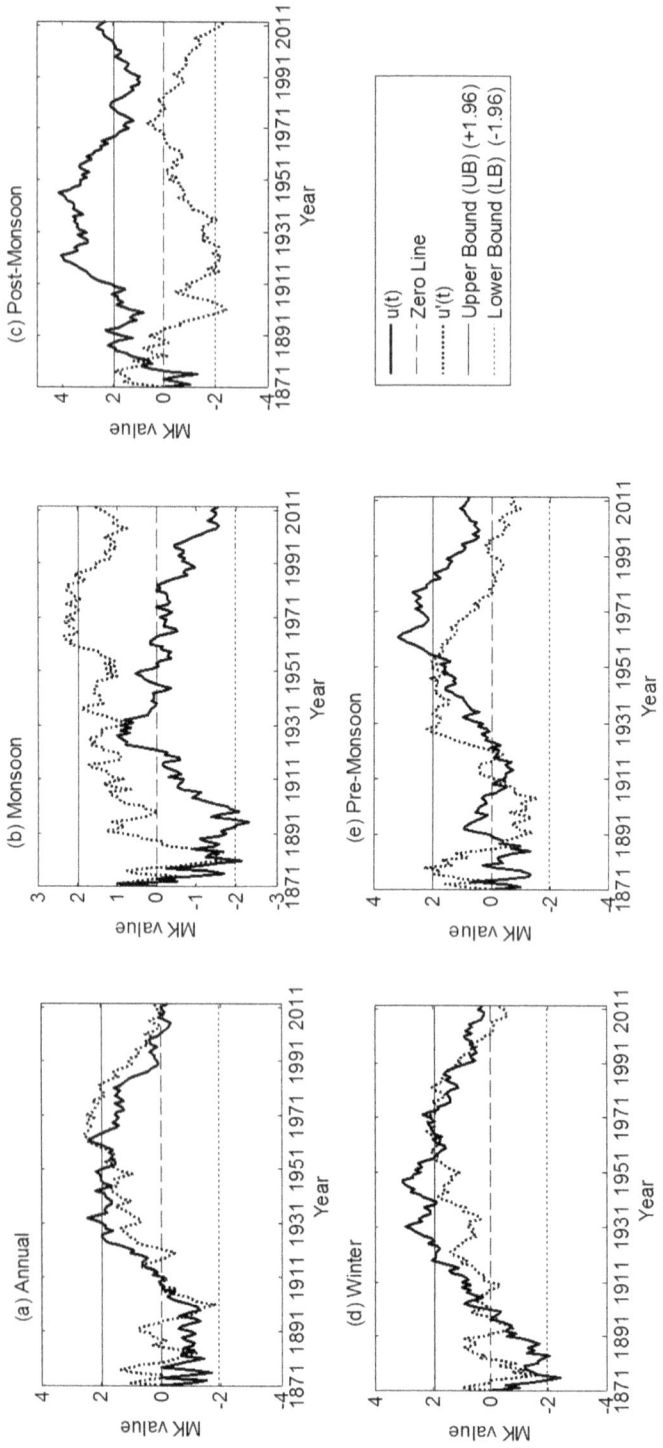

FIGURE 4.4 Sequential MK values for annual and seasonal rainfalls in Kerala sub-division (a) annual; (b) monsoon; (c) postmonsoon; (d) winter; (e) premonsoon seasons.

significant trends for relatively longer time spells; (ii) the early commencement of an apparent trend was observed for post-monsoon rainfall of Kerala subdivision.

It is well understood that the meteorological time series often possess nonlinear trend (Wu et al. 2007). The decomposition based methods are reported to possess some advantages over the MK and SS methods, as they can give the specific shape of inherent nonlinear trend and hence can give some information on the time of occurrence of changes (increasing or decreasing) over different time windows within a time series (Sang et al. 2014; Unnikrishnan and Jothiprakash, 2015). Therefore, this study applied EMD for extraction of nonlinear trend of the annual rainfall time series, different seasonal series and time series of rainfall of different months from Kerala meteorological subdivision. The results of EMD-based trend analysis of rainfall time series of different months is presented in Figure 4.5. The figure shows that monotonic nature of trend prevails for January, February, July, September and October, and there are differences in the year in which a character change occurs (increasing to decreasing or vice versa) for rest of the months.

The EMD based trend analysis and its Fourier spectrum for the different seasonal series are presented in Figure 4.6, which reveals that only smaller frequency intervals have higher Power Spectral Density (PSD) values and for the rest it is almost zero. This implies that the trend infer extracted are corresponding to low-frequency fluctuations. Also, to investigate the statistical significance of trend, the test proposed by Wu and Huang (2004b) is applied. In this test, the significance line of white noise series is considered as a reference and the IMFs, whose energy level falls above the upper confidence line can be considered as significant. Generally, the first IMF is considered as the reference IMF for finding the normalized energy to prepare a plot between log(Mean Normalized Energy) and log (Mean Period) of different IMFs. The statistical significance test (SST) of modes of annual and seasonal rainfall series of Kerala is performed and presented in Figure 4.7. As the energy level of 'residue' is above the upper significance line, the nonlinear trends of these series can be considered to be significant (Sang et al. 2014).

4.4 HILBERT HUANG TRANSFORM ANALYSIS OF RAINFALL TIME SERIES

The trend analysis study gave broad insights on the rainfall in India along with the role of climatic oscillations of inter annual periodicity on the rainfall variability. Multiscale spectral analysis may give better insights on the periodic properties of the rainfall, and it may help in finding the links of rainfall with climatic oscillations quantitatively and eventually help in improved predictions of rainfall. In the next section, the HHT analysis of monsoon rainfall from AI and five nonoverlapping macro regions is presented.

4.4.1 HILBERT HUANG TRANSFORM ANALYSIS OF MONSOON RAINFALL OVER INDIA

In the HHT analysis of ISMR, first the All India Summer Monsoon Rainfall (AISMR) time series is decomposed into suitable number of Intrinsic Mode Functions (IMFs)

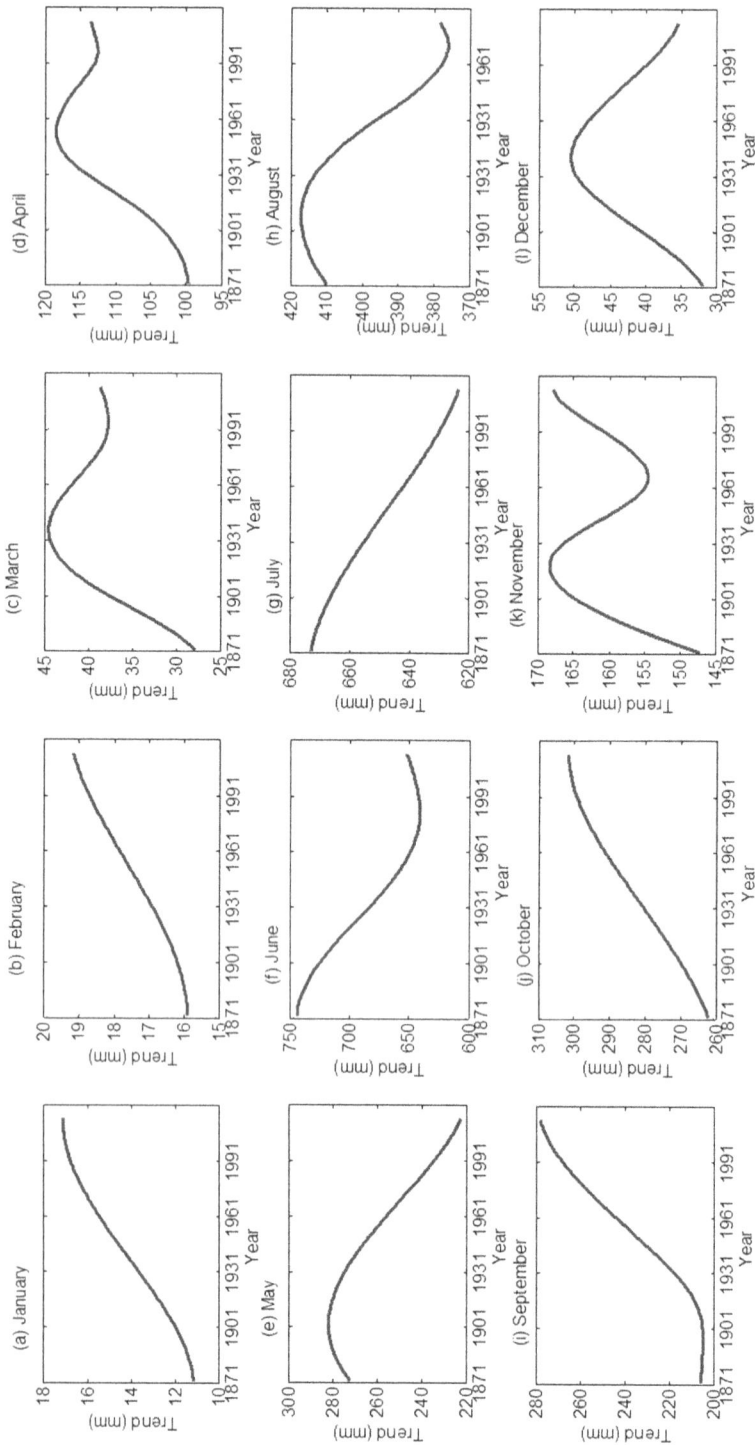

FIGURE 4.5 Nonlinear trend extracted for different month's rainfall time series of Kerala (1871 to 2012) using EMD.

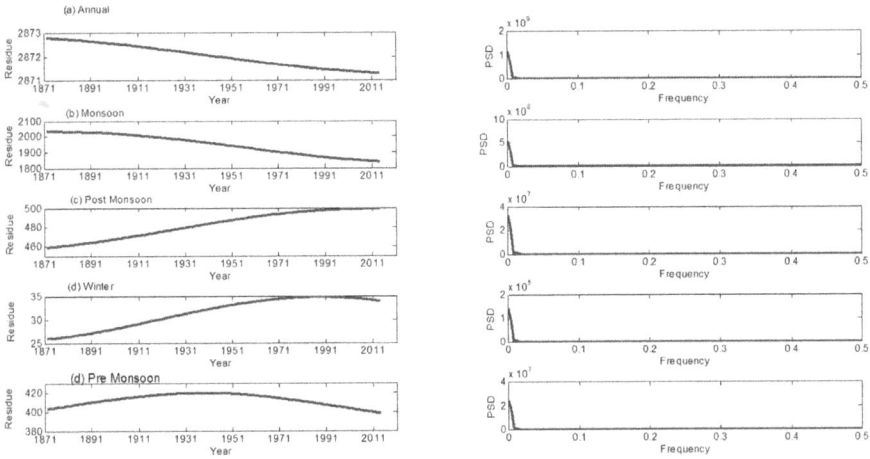

FIGURE 4.6   Trend components of annual and seasonal rainfalls (left panels) of Kerala and corresponding Fourier spectrums (right panels).

and residue by the EMD and two of its variants namely EEMD and CEEMDAN. For running the EEMD and CEEMDAN algorithms, a noise standard deviation of 0.2 and 500 realizations were used to get the ensemble of modes as per the past studies (Torres et al. 2011; Antico et al. 2014). The results of EMD, EEMD and CEEMDAN based decomposition are presented in Figure 4.8. The number of sifting iterations taken for evolving different IMFs by the EEMD and CEEMDAN methods are shown in box and whisker plot in Figure 4.9. The plots show that number of sifting iterations required to evolve each mode is relatively less for CEEMDAN as compared to EEMD.

The probability density functions (PDFs) of first five IMFs generated by the three methods of decomposition for AISMR data is shown in Figure 4.10. The deviation from normal distribution is more for higher modes and all the methods show multimodality behavior in decomposition for the higher modes. This is because, in the higher modes, the IMFs contain a smaller number of oscillations; therefore, the distribution becomes less smooth. For a larger sample size, the IMFs of the higher order modes will have more oscillations and the distribution will follow the normal distribution according to the central limit theorem (Papoulis 1986).

The correlation of IMFs by three methods with the original dataset is shown in Figure 4.11(a). It shows highest correlation for the first IMF by all the three methods. For the lower modes all three methods give similar value of coefficient of correlation, but for the higher modes the extraction of residues shows slight difference by the three methods of decomposition. The EMD and CEEMDAN methods decompose the data into six different frequency scales, while the EEMD extracts seven frequency scales from the dataset. But the last two modes are found to be showing the characteristics of the residue as those modes are not oscillatory about zero. The decomposed modes from a time series should be orthogonal to each other and the

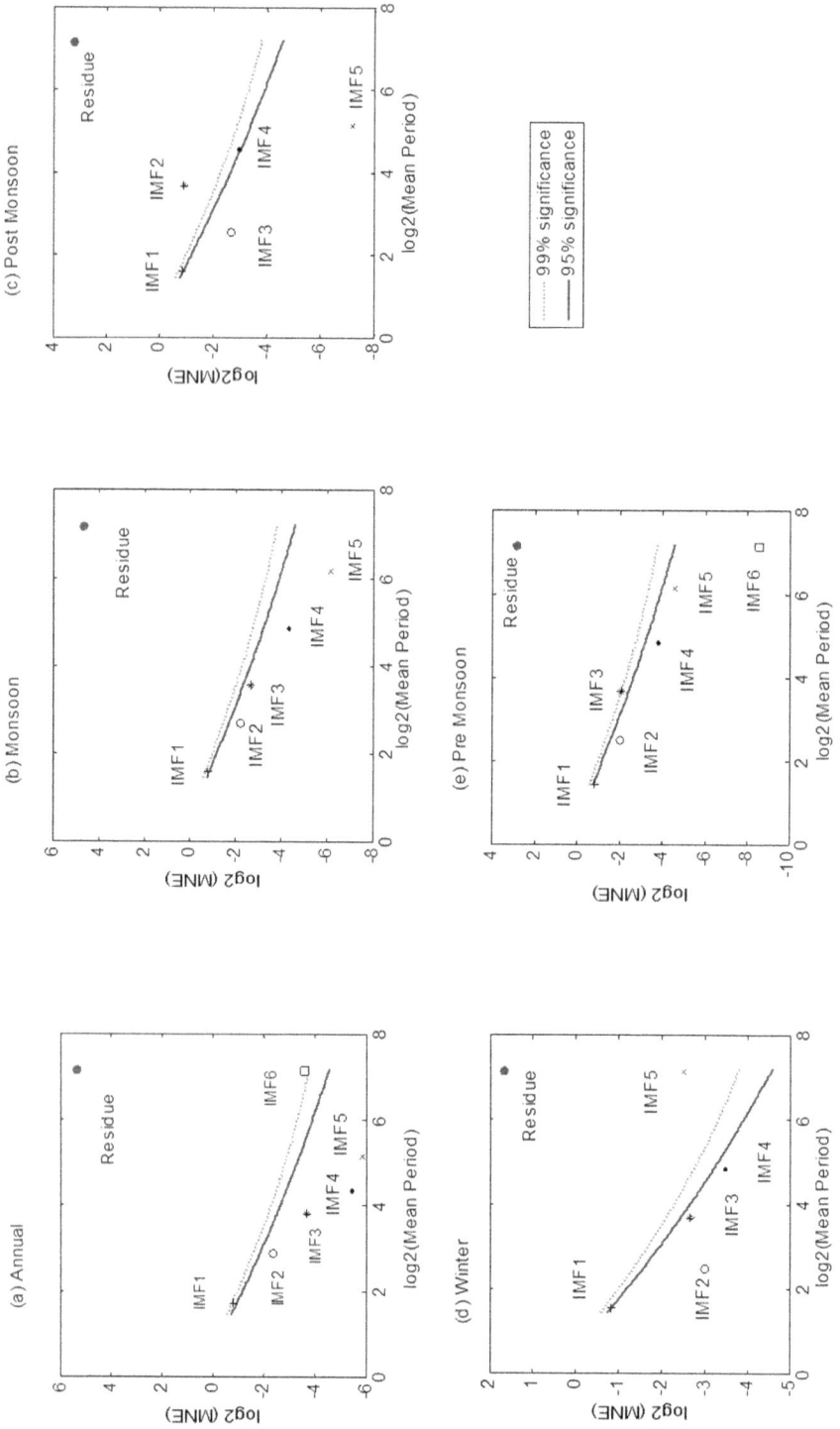

FIGURE 4.7   Statistical significance test of modes of annual and seasonal rainfalls in Kerala.

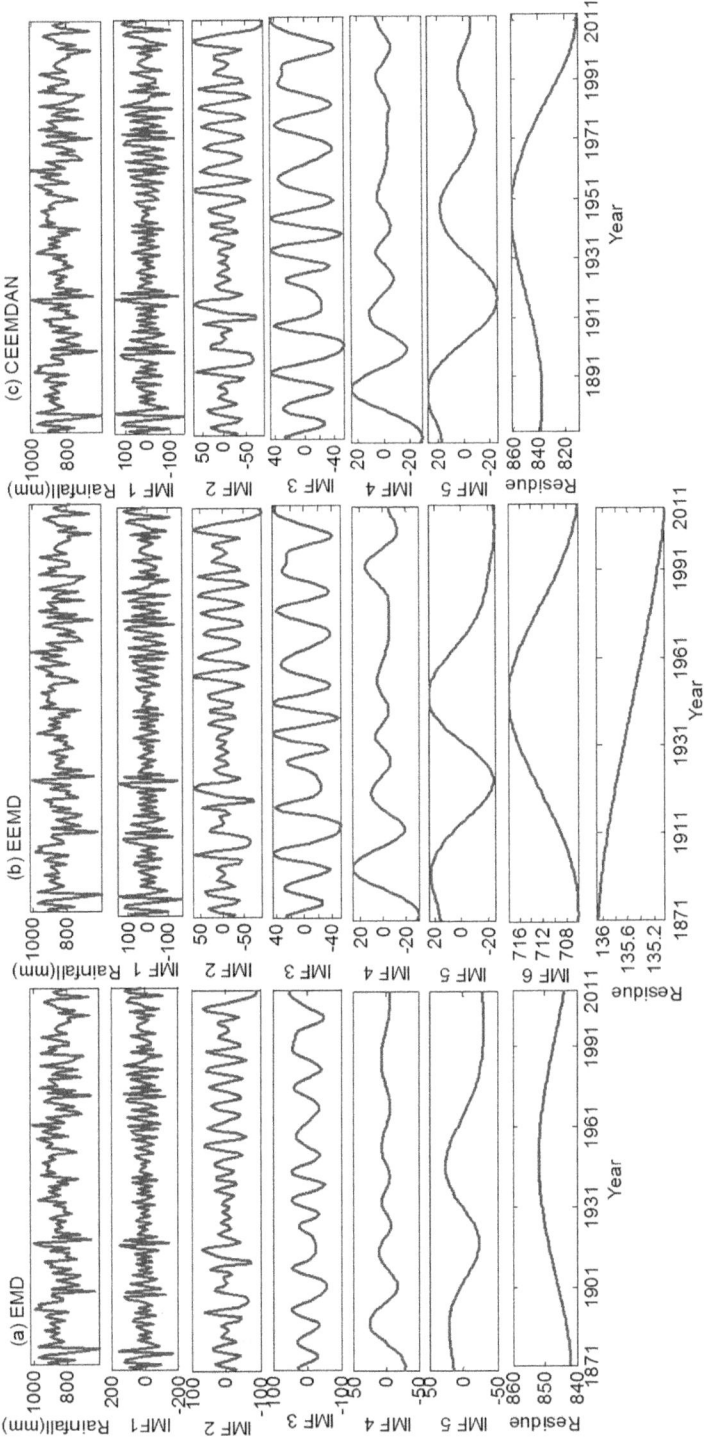

FIGURE 4.8 Decomposition of AISMR time series using EMD, EEMD and CEEMDAN methods.

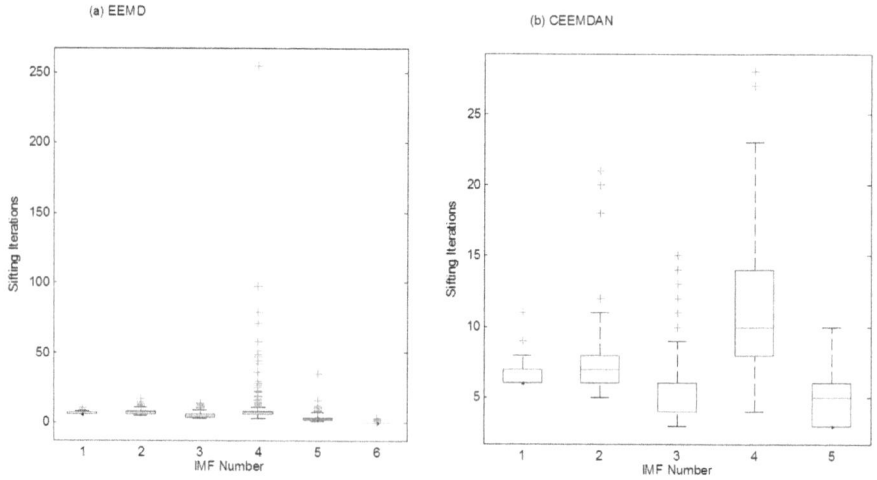

FIGURE 4.9 Box and Whisker plots of number of sifting iterations for generating different modes (excluding the residue) by EEMD and CEEMDAN methods.

decomposed components are tested for orthogonality using the following expression (Huang et al. 1998) called the Index of Orthogonality (*IO*)

$$IO = \frac{1}{T}\sum_{t}\frac{1}{[X(t)]^2}\left(\sum_{l=1}^{n}\sum_{m=1}^{n}IMF_l(t)IMF_m(t)\right) \tag{4.1}$$

where *l* and *m* are the indices for the IMFs.

A lower value of *IO* indicates the closeness to orthogonal behavior. A set of perfect orthogonal IMF components will give zero but in practice, the value of *IO* smaller than 0.1 is acceptable (Wu et al. 2011). The *IO* values are found to be 0.000218, 0.00247 and 0.000197 for EMD-, EEMD- and CEEMDAN-based decompositions, respectively. The values show that the overall orthogonality property is less for the EEMD-based decomposition when compared with EMD and CEEMDAN-based decomposition. This may be because the ensemble averaged modes may not always be satisfying the properties of an IMF and when such IMFs are subjected to Hilbert Transform, they may lead to negative values of instantaneous frequency which lacks physical meaning. The different frequency scales obtained by the decomposition represents the variability in the dataset. The ratio of variance of an IMF to the variance of the data series (in %) (RoV), estimated by three methods are presented in Table 4.4. The mean period of the time series is calculated by counting the number of extrema points divided by the number of data samples (Barnhart and Eichinger 2011b).

Table 4.4 shows that the mean period of most of IMFs is roughly twice the period of previous IMF. Because the IMFs are nearly dyadic, each IMF is the sum of all frequency (periodic) scales within the dyadic range of that IMF, i.e., there is no mixing of drastically different modes. A plot between log2(*T*) for different IMFs is shown

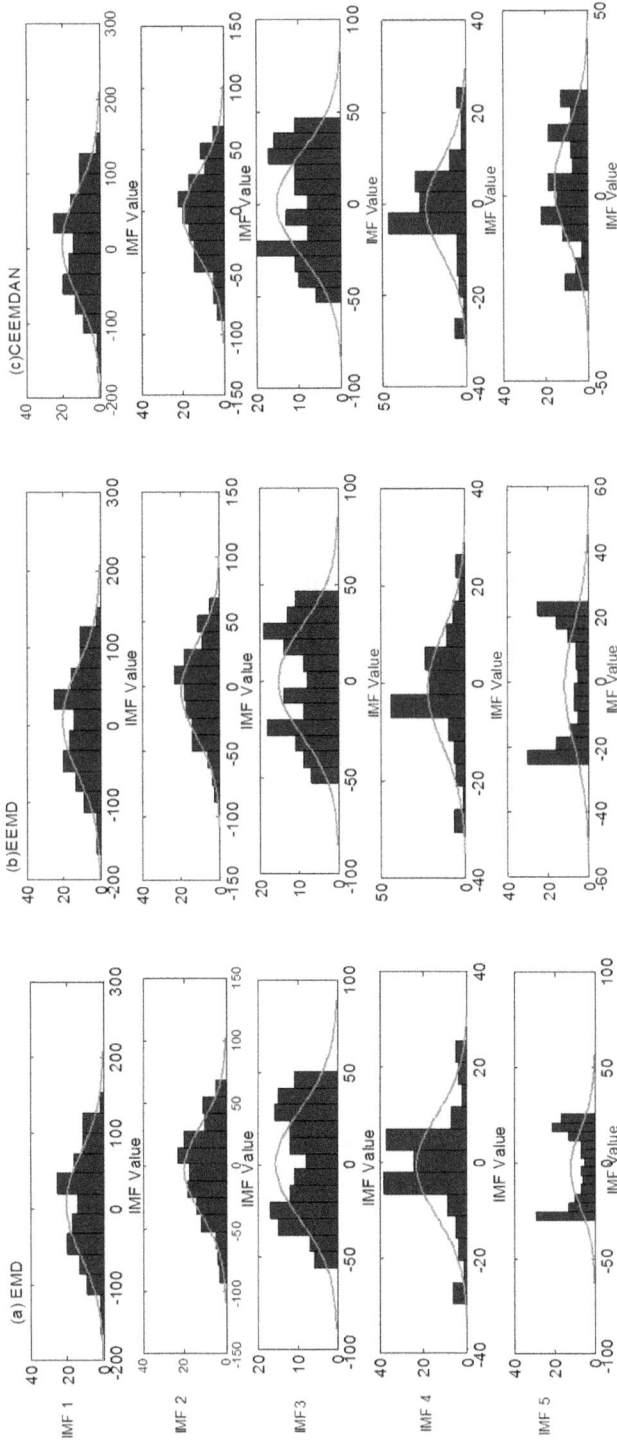

FIGURE 4.10 PDFs of first five IMF components of AISMR time series, obtained by the EMD, EEMD and CEEMDAN methods. The normal distribution plots are superposed.

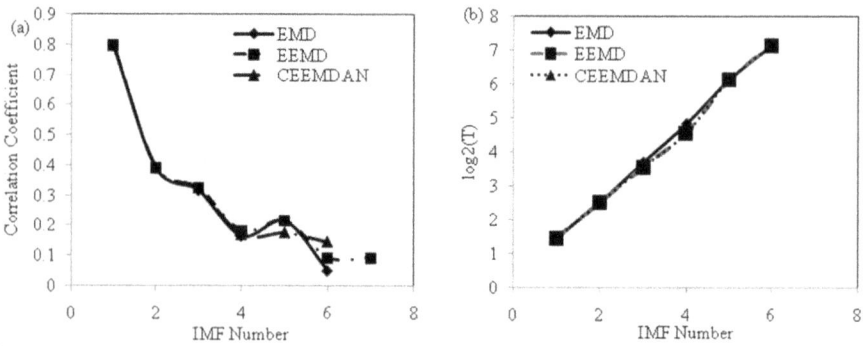

FIGURE 4.11   (a) Comparison of correlation coefficient of IMFs obtained by EMD, EEMD and CEEMDAN with the AISMR dataset (b) Near dyadic nature of IMFs obtained by EMD, EEMD and CEEMDAN

TABLE 4.4
Comparison of Ratio of Variances (RoV) and mean period of IMFs of AISMR data obtained using EMD, EEMD and CEEMDAN methods

IMF No	EMD		EEMD		CEEMDAN	
	RoV	Mean period	RoV	Mean period	RoV	Mean period
1	66.015	2.73	66.027	2.731	66.081	2.731
2	17.938	5.68	17.541	5.680	17.500	5.680
3	12.760	12.91	11.480	11.833	11.819	11.833
4	1.584	28.4	1.8356	23.667	1.498	23.667
5	5.772	71.0	4.776	71.0	3.356	71.0
6	0.146	142.0	0.245	142.0	1.927	142.0
7	—	—	0.002	142.0	—	—

in Figure 4.11(b). A perfectly linear fit can be considered as decomposition with no 'mode mixing' (Barnhart 2011). Figure 4.11(b) indicates that the three plots show nearly dyadic filtering of given dataset with a marginal deviation for EMD than the two ensemble methods for the modes 3 and 4. Hence it is perceptible in the decomposition for the used dataset. Because the extracted modes are showing better orthogonality character (i.e., low value for index of orthogonality) and require fewer sifting iterations for CEEMDAN than that for EEMD, hence the CEEMDAN is used for decomposition of all the remaining five monsoon rainfall time series considered in this study.

The mean periods obtained by the EEMD and CEEMDAN (Table 4.4) show a close matching with those reported in earlier studies (Narasimha and Kailas 2001; Iyengar and Raghu Kanth 2005). The monsoon rainfall of the five nonoverlapping regions are also decomposed using the CEEMDAN and resulting modes of are shown in Figure 4.12.

In these decompositions, the last IMF is the residue which is an indication of trend in the dataset. The rainfall series of North West India and Peninsular India are showing an increasing trend in the recent past. Identical observations were also made by Iyengar and Raghu Kanth (2005). The remaining series are showing a reducing trend in rainfall unlike those observed in the past studies by Iyengar and Raghu Kanth (2005). As the updating of the dataset for more than 20 years is made in this study when compared with the past study, so it can be inferred that the observed reverse trend may be due to the reduction of the rainfall in the recent past in these regions (an evidence of change in climate). The mean periodicities of different IMFs for all the regions are shown in Table 4.5. At least five different periodicities (each associated with one IMF) are observed in all the six rainfall time series considered in this study. Furthermore, the Morlet wavelet power spectrum of AISMR and rainfall of different nonoverlapping macro regions is prepared and the mean periodicity obtained by the CEEMDAN is superimposed upon the Morlet wavelet spectrum as shown in Figure 4.13.

Table 4.5 shows that IMF1 possess an average period of 2.6–3.02 years in different regions. The periodicities in different modes of ISMR and the periodicity of natural climate forcing infer the preliminary evidence of possible teleconnections of ISMR with different global climate indices. The periodicity of QBO is 2–3 years and observing such a mode in ISMR time series suggests the possible link between the two (Chattopadhyay and Bhatla 2002; Claud and Pascal 2007). The ENSO has 3–7 years' periodicity (Kripalani and Kulkarni 1997a, 1997b; Pokhrel et al., 2012) and the average period of second mode obtained by the decomposition of ISMR from different region vary between 5.68 and 6.17. Hence a possible link of ISMR and ENSO could be assessed. The IMF3 has a mean periodicity between 10.9 and 11.9 years in all regions except NWI. It is well understood that sunspot time series have a mean periodicity of 11 years (Usoskin and Mursula 2003; Barnhart and Eichinger 2011a); hence a preliminary notion on the link between ISMR and sunspot cycle could be established. The mean period of fourth mode (IMF4) is found to be between 23–29 years, which could be possibly linked with the tidal forcing of similar periodicity (Campbell et al. 1983; Iyengar and Raghu Kanth 2005). Campbell et al. (1983) examined the precipitation records of northern India for the period 1895–1975 using the Eigen vector analysis. By comparing its dominant frequencies of the precipitation data with that of soli-lunar tidal potential at the latitude of northern India they hypothesized that the tidal effects modulates the advance of monsoon front and proposed a method for prediction of June rainfall one year in advance using the information on tidal frequencies. The IMFs having multi decadal periodicity of more than 60 years (IMF5) may represent a possible association of AMO with the Indian monsoon rainfall (Goswami et al. 2006; Lu et al. 2006). Moreover, to explain the link between climate indices and the different modes obtained by the multiscale decomposition, a plot between the IMF and the respective climate index time series can be plotted. As the annual datasets for the period 1871–2012 are available only for the sunspot number (SN) time series and AMO time series, the plot of IMF3 and IMF5 only are considered. The comparison of respective IMF and climate index time series

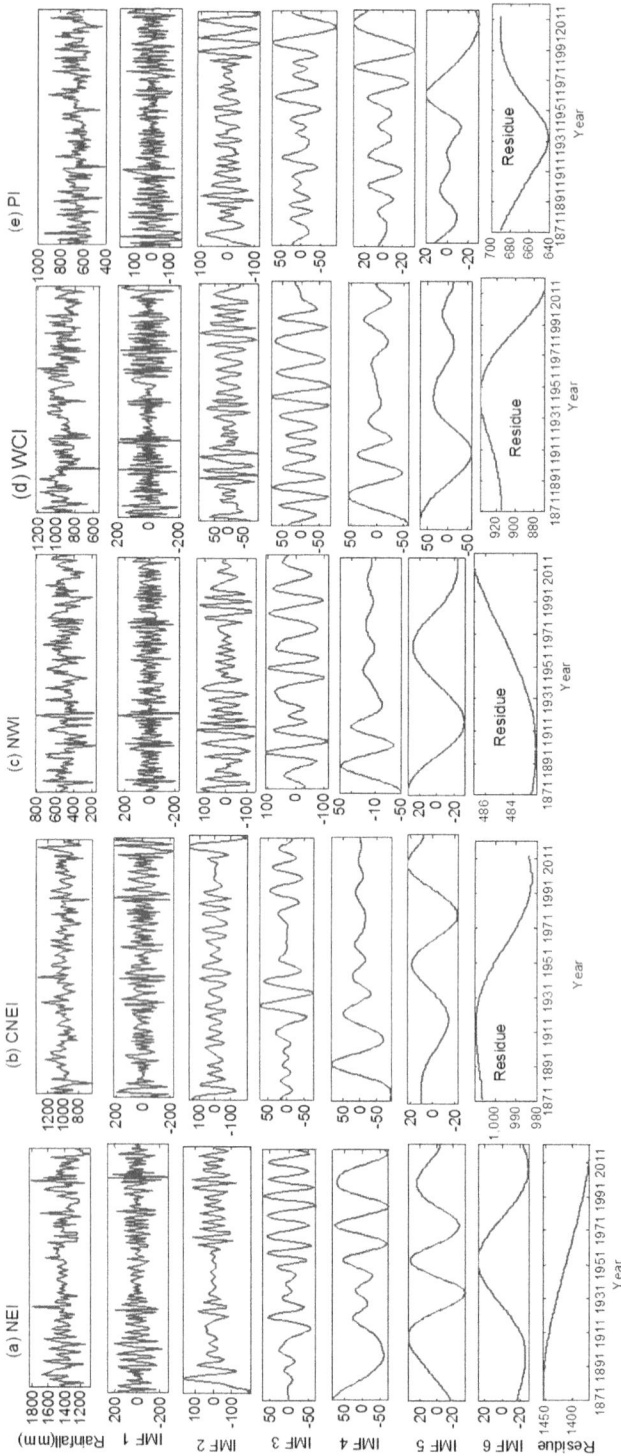

FIGURE 4.12 (a)–(e) Orthogonal modes of rainfall time series for five nonoverlapping macro regions in India.

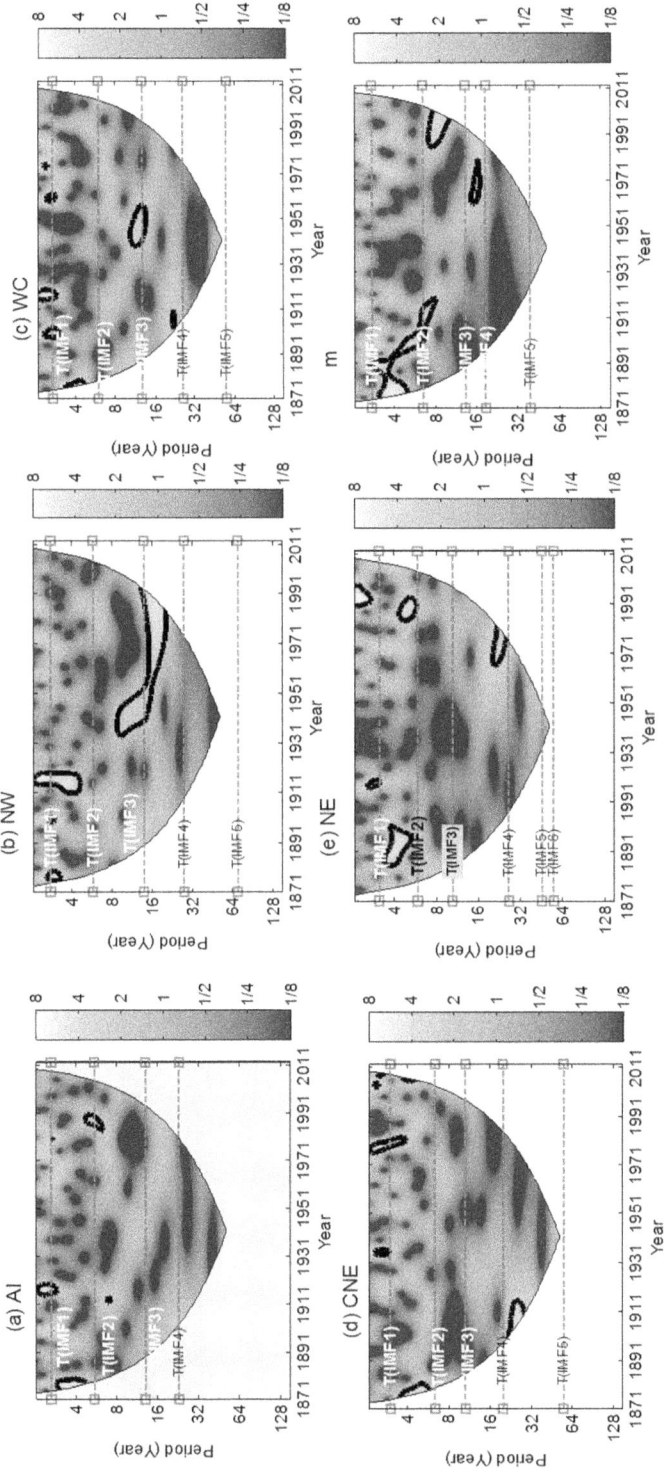

FIGURE 4.13 Morlet wavelet spectrum of AISMR and monsoon rainfall time series from different macro regions with periods of IMF's superimposed.

TABLE 4.5
Mean period (T in years) and percentage variability explained (VE) by different IMFs of rainfall time series from different regions of India

Mode Number	NEI		CNEI		NWI		WCI		PI	
	T	VE (%)	T	VE (%)	T	VE (%)	T	VE (%)	T	VE (%)
IMF1	2.9	55.9	3.0	62.01	2.8	54.78	2.6	60.96	2.73	63.16
IMF2	5.7	19.5	6.4	23.15	5.7	23.55	5.7	13.63	6.17	21.29
IMF3	10.9	5.52	10.1	4.44	14.2	17.05	11.8	11.87	11.83	7.80
IMF4	28.4	10.7	20.3	8.71	28.4	1.95	28.4	3.75	20.28	2.13
IMF5	47.3	2.09	47.3	1.02	71.0	2.66	47.3	7.39	47.33	2.18
IMF6/Residue	71.0	1.94	71.0	0.67	142.0	0.01	142.0	2.40	142.0	3.45
Residue	142.0	4.33								

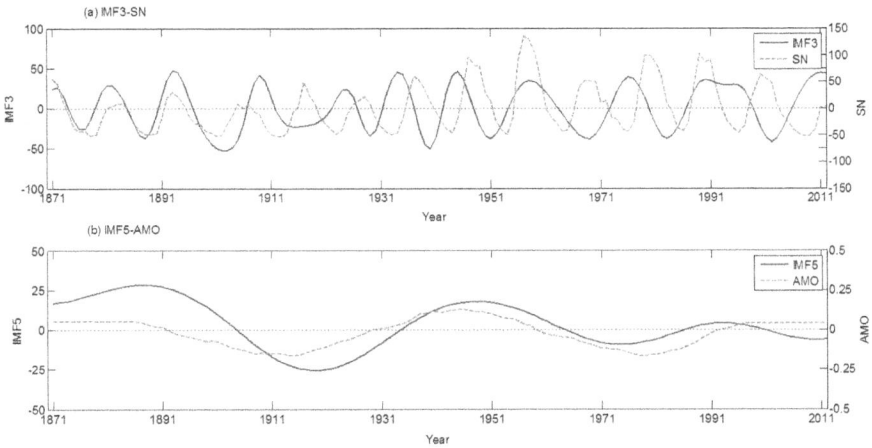

FIGURE 4.14 Comparison of IMF of rainfall and the zero mean climate index time series (a) IMF3 with the SN time series; (b) IMF5 with the AMO time series. (From Adarsh 2017).

are presented in Figure 4.14. The visual comparison between the IMF and time series show striking similarity and similar evolution in most of the time duration. This further confirms the distinct possibility of the link of SN and AMO with the ISMR.

The IMFs are the appropriate components to perform the HT and which may give instantaneous frequencies and amplitudes. To avoid the possibility of getting instantaneous frequencies and amplitudes of less physical meaning (such as negative frequency) and mathematical incorrectness the NHT-DQ scheme is invoked for the HT analysis. The instantaneous frequency trajectories of different time series are shown in Figure 4.15. The statistical properties of instantaneous amplitudes and frequencies are given in Table 4.6 and Table 4.7, respectively. The integration of Hilbert spectra of all IMFs of a time series over the time will give the marginal spectrum ($h(\omega)$), which gives the amplitude of considered time series (say, rainfall) as a function of frequency. It represents the cumulative amplitude over the entire data span in a probabilistic sense, i.e., the existence of amplitude at a frequency ω means there is a higher likelihood for such amplitude to occur at that frequency. It is helpful to find the dominant periodicities in a time series. The marginal Hilbert spectrum (MHS) of rainfall time series from All India and the five nonoverlapping regions are presented in Figure 4.16.

From the spectral analysis of summer monsoon rainfall time series data pertaining to six different regions, it is noted that the high amplitude rainfall in all regions is due to the first few IMFs and the concentration of highest amplitudes is at mean frequency of ~ 0.40 cycles/year (period of 2.5 years). This can be inferred as the dominant frequency. But it can be noted that the dominant frequency varies with time. This shows the influence of natural cycles of short-term periodicity (< 5 years) in high rainfall events. The time-frequency representation of different modes can be distinguished one from other and the scattered points in the color map have established the intermittency of the processes involved. The discontinuity in the frequency is observed typically for high frequency modes (IMF1-IMF3). On the other hand, continuous

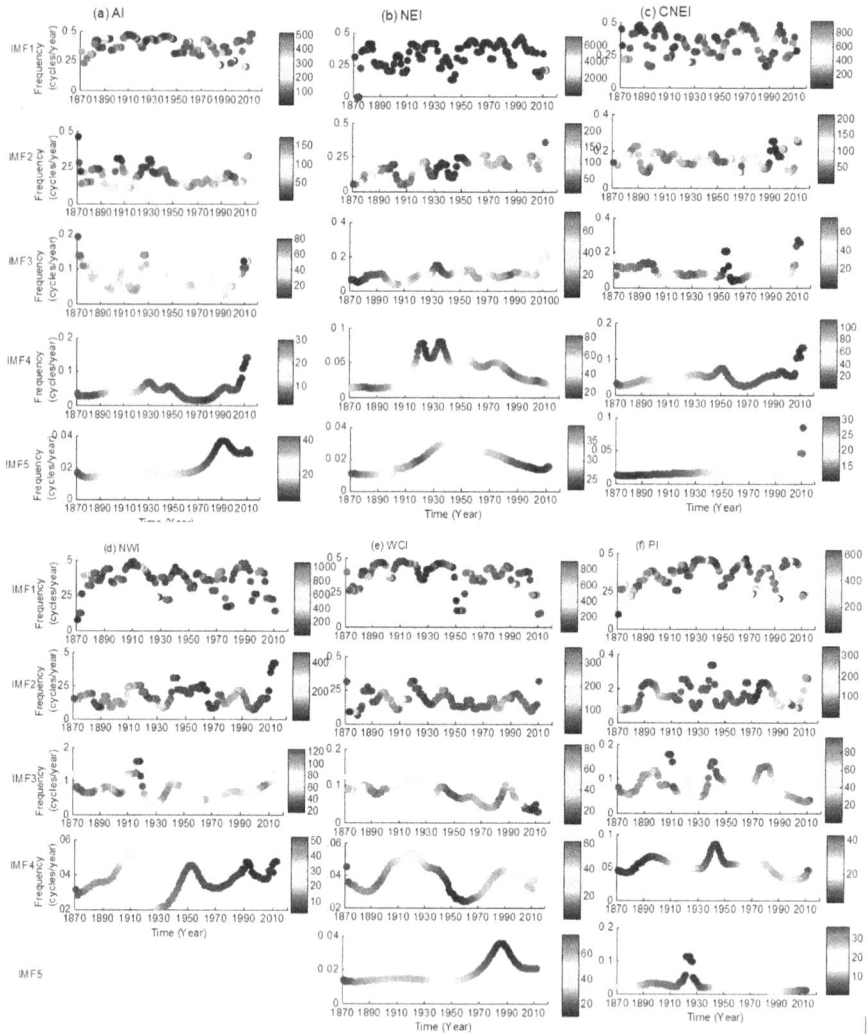

FIGURE 4.15 Instantaneous frequency trajectories from the HHT applied to each IMF component of monsoon rainfall for the different regions: (a) All India (AI), (b) North East India (NEI), (c) Central North East India, (d) North West India, (e) Western Central India (WCI), (f) Peninsular India (PI). The color scale indicates the amplitudes in mm.

regions of color are observed for frequencies of higher order modes. The intermittency of instantaneous frequencies in the lower order modes shows the nonlinearity of the associated physical processes (Franceschini and Tsai 2010). The range of frequency is quite high for the lower oscillatory mode and the standard deviation of frequency is also quite high for (IMF1 in particular). This high modulation of frequency for lower intrinsic modes depicts the nonstationarity of the processes involved. The time-frequency-amplitude plots depict the temporal change in magnitude of rainfall

TABLE 4.6
Mean amplitude (MA), maximum amplitude (Max A) and standard deviation of amplitude (Std. A) of rainfall time series from different regions

IMF No.	AI			NEI			CNEI		
	MA	Max A	Std. A	MA	Max A	Std. A	MA	Max A	Std. A
IMF1	101.963	520.842	75.859	250.651	7139.700	729.475	143.119	983.487	118.94
IMF2	44.238	172.955	25.140	69.510	225.000	46.692	69.251	213.840	39.807
IMF3	38.993	80.852	11.936	39.277	78.400	22.216	30.528	76.214	22.190
IMF4	10.892	29.876	8.681	59.344	82.700	21.222	38.947	113.172	32.703
IMF5	19.025	42.892	10.230	31.167	39.100	5.167	17.736	31.037	4.977
IMF No.	**NWI**			**WCI**			**PI**		
IMF1	165.990	1069.90	147.329	145.075	912.366	123.831	133.659	642.186	102.600
IMF2	85.318	474.200	53.525	63.572	363.816	37.400	62.456	346.546	39.350
IMF3	74.782	125.100	25.377	57.049	83.769	19.939	39.977	98.025	23.825
IMF4	19.785	52.800	16.355	28.462	61.511	17.511	18.152	42.676	10.749
IMF5				40.741	75.806	20.086	15.568	36.577	9.182

due to the climate oscillations of different scales. In the recent past the IMF4 of NEI time series and IMF3-IMF5 of the Peninsular India (PI) show a concentration of higher amplitude (i.e., the associated cycles/processes have more influence on the rainfall events in these regions). The concentration of higher amplitudes due to the processes of low frequency (IMF5) are noticeable in the first few decades of observation period (1870–1920) except for Peninsular India (PI) and the Central North East India (CNEI) region.

From the marginal spectrums of different regions, it is observed that, there are multiple prominent peaks for the AI and PI time series and concentration of such peaks are at frequency greater than 0.35. Except at Central North East and North East India, the events of highest amplitude are expected at a dominant frequency of ~0.45cycles/year. In NE region, the extreme rare events (of considerably high magnitude) are expected with a frequency of 0.25 cycles/year. Excluding such rare events, the response is similar to that in other regions. The marginal spectrum further shows the multiple high amplitude events in all regions with a frequency of 0.25–0.3 cycles /year. Also in peninsular region, multiple high rainfall events (say > 600 mm) are expected with varying time scale.

4.4.2 HHT Analysis of Rainfall at Subdivisional Scale

The spatial variability of rainfall pattern in India has huge impact in the agrarian economy of the country. The climate of the Himalayan range in India is sub-freezing, while the coastal regions of India experience a tropical climate. Lack of consistent patterns stresses the need for a more detailed regional analysis to explore the changes

TABLE 4.7

Frequency range, mean frequency (Mean F) and standard deviation of frequency (Std. F) of rainfall time series from different regions

IMF No.	AI			NEI			CNEI		
	Frequency Range	Mean F	Std. F	Frequency Range	Mean F	Std. F	Frequency Range	Mean F	Std. F
IMF1	0.000-.480	0.371	0.092	0.000-0.472	0.321	0.115	0.165-0.486	0.338	0.093
IMF2	0.000-0.465	0.181	0.082	0.000-0.365	0.171	0.079	0.085-0.265	0.157	0.041
IMF3	0.027-0.192	0.076	0.028	0.041-0.242	0.096	0.032	0.036-0.273	0.094	0.042
IMF4	0.015-0.142	0.043	0.022	0.014-0.080	0.038	0.020	0.026-0.132	0.047	0.020
IMF5	0.014-0.037	0.020	0.007	0.011-0.030	0.019	0.007	0.012-0.085	0.018	0.009

IMF No.	NWI			WCI			PI		
	Frequency Range	Mean F	Std. F	Frequency Range	Mean F	Std. F	Frequency Range	Mean F	Std. F
IMF1	0.079-0.488	0.359	0.088	0.000-0.476	0.372	0.114	0.095-0.467	0.359	0.079
IMF2	0.069-0.419	0.172	0.068	0.000-0.320	0.161	0.101	0.076-0.338	0.161	0.061
IMF3	0.033-0.159	0.072	0.025	0.030-0.130	0.079	0.023	0.034-0.173	0.083	0.033
IMF4	0.021-0.053	0.036	0.008	0.024-0.054	0.038	0.008	0.033-0.086	0.052	0.012
IMF5				0.013-0.036	0.018	0.006	0.011-0.115	0.025	0.018

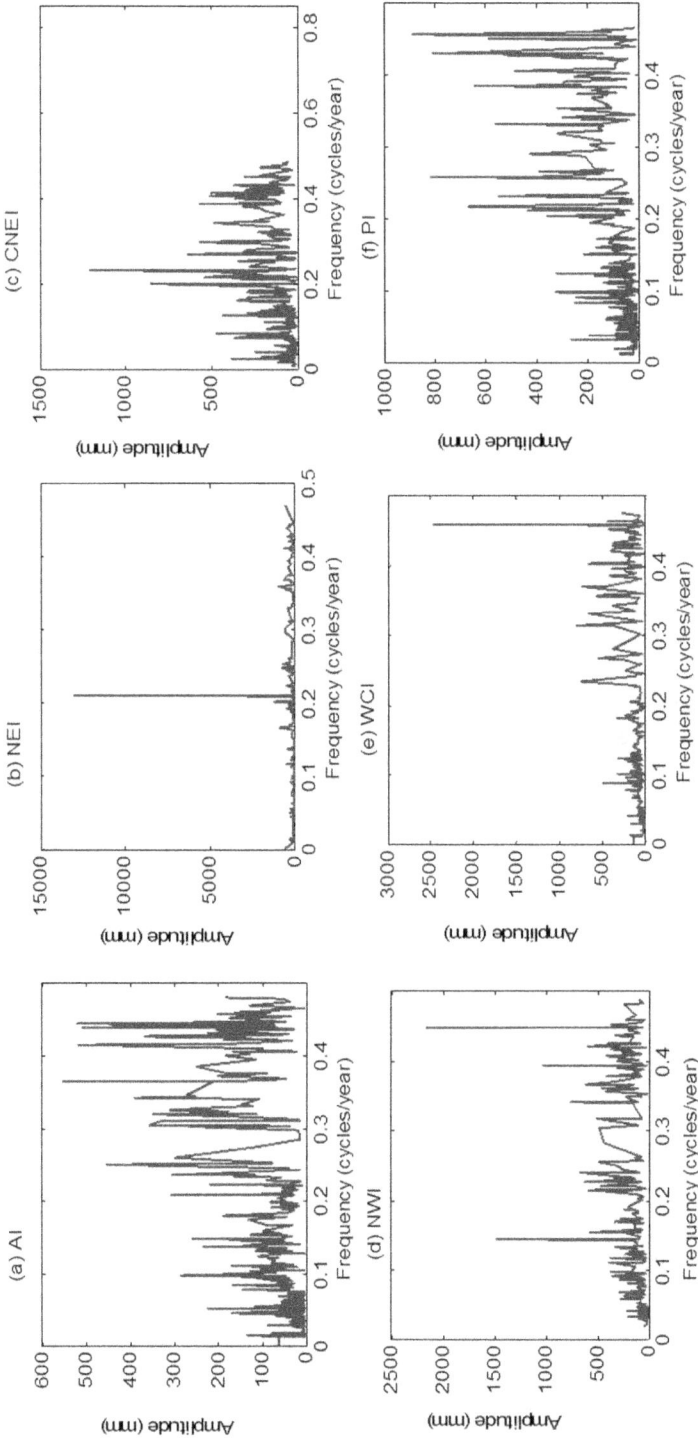

FIGURE 4.16 (a)–(h) Marginal Hilbert spectrum of monsoon rainfall time series for different macro regions in India.

in hydroclimatic variables (Panda et al. 2013). The area weighted monthly rainfall data prepared by IITM Pune (www.tropmet.res.in) for the period 1871–2012 for different meteorological subdivisions in India is useful for such regional analysis. In this section, the rainfalls of monsoon and post-/premonsoon seasons at Kerala are analyzed to demonstrate the usefulness of HHT.

The monsoon and premonsoon rainfall of Kerala are decomposed into different IMFs using the CEEMDAN. In CEEMDAN a standard deviation of 0.02 and 500 realizations were invoked to get the ensemble of modes considering suggestion from the past studies (Torres et al. 2011; Antico et al. 2014). Subsequently, the obtained IMFs except the residue are subjected to the Hilbert Spectral Analysis (HSA) to perform a detailed time-frequency characterization of all the eight rainfall time series by employing NHT-DQ scheme. The Hilbert spectrum of each IMFs are presented as instantaneous frequency trajectories in which the y-axis refers to the instantaneous frequency and the color scale depict the amplitude (magnitude of rainfall values). The IMF components and the instantaneous frequency trajectories of the monsoon and post monsoon rainfall of Kerala subdivision is presented in Figure 4.17. In this analysis the residue appear as a single peaked curve with the peak ~1930. During ~1871–1930 the postmonsoon rainfall show an overall increase of ~150 mm, but afterward (1930–2012) a marginal decreasing trend is noticed. Furthermore, the periodicity of the modes estimated by a zero crossing method (Huang et al. 2009b) and the statistical significance of IMFs is determined by comparing the energy level of different IMFs with the spread function of white noise (Wu and Huang 2004a, 2004b).

The instantaneous frequency trajectories of monsoon and postmonsoon rainfall are integrated to get the MHS of the respective series. The periodicity and linear correlation with the respective time series in Kerala subdivision is presented in Table 4.8, while the results of statistical significance test and MHS are provided in Figure 4.18.

From Table 4.8, the highest correlation is observed for the first IMF, for both the time series. The correlation is similar to those reported in a regional scale study by Iyengar and Raghu Kanth (2005) and it enables the possibility of statistical forecasting of seasonal rainfall in the subdivision. At least five meaningful periodic scales could be identified from the decomposition while the last few modes are indicating long-term climate variability. Here also, different periods can be associated with the climatic oscillations such as QBO, ENSO, sunspot cycle, tidal forcing and AMO and last component refers the slowly varying climate oscillations. The final residue shows a monotonic decreasing trend in the monsoon rainfall of Kerala subdivision and a statistically significant increasing trend is noticed for the postmonsoon rainfall, which is consistent with recent studies (Krishnakumar et al. 2009). The results of statistical significance test of IMF components and the MHS obtained by integration of Hilbert spectra of monsoon and postmonsoon rainfall are presented in Figure 4.18.

In the monsoon rainfall series, the statistical significance of different IMFs is similar to the results reported by Iyengar and Raghu Kanth (2005) for the regional scale analysis. Interestingly, for the postmonsoon rainfall in Kerala subdivision, the IMF3 of Kerala subdivision (having near decadal periodicity) is found to be statistically significant. The study by Adarsh and Janga Reddy (2015) on trend analysis of postmonsoon rainfall in Kerala using the DWT also stated a possible association

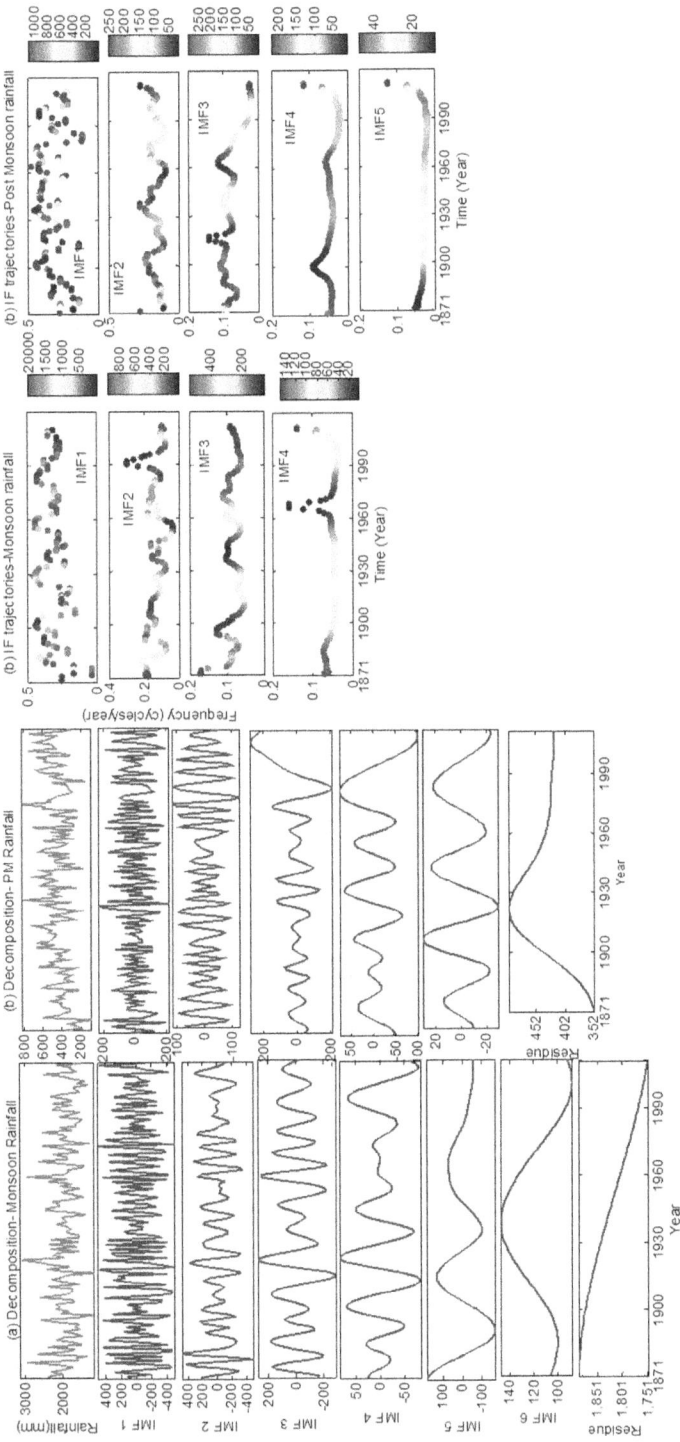

FIGURE 4.17 The decomposition of monsoon and post monsoon rainfall of Kerala along with the instantaneous frequency trajectories of the IMF components. The color scale indicates the amplitudes in mm.

TABLE 4.8
Correlation of IMFs with the time series and the mean period of IMFs of monsoon rainfall and postmonsoon (PM) rainfall of Kerala subdivision

IMF Number	M		PM	
	R	T	R	T
IMF1	0.704	2.958	0.676	3.021
IMF2	0.515	6.455	0.285	5.917
IMF3	0.325	11.833	0.410	12.909
IMF4	0.219	17.750	0.017	23.667
IMF5	0.219	71.000	0.047	35.500
IMF6	0.133	142.000	0.263	142.000
IMF7	0.134	142.000		

Note: R-correlation coefficient; T-Mean period in year.

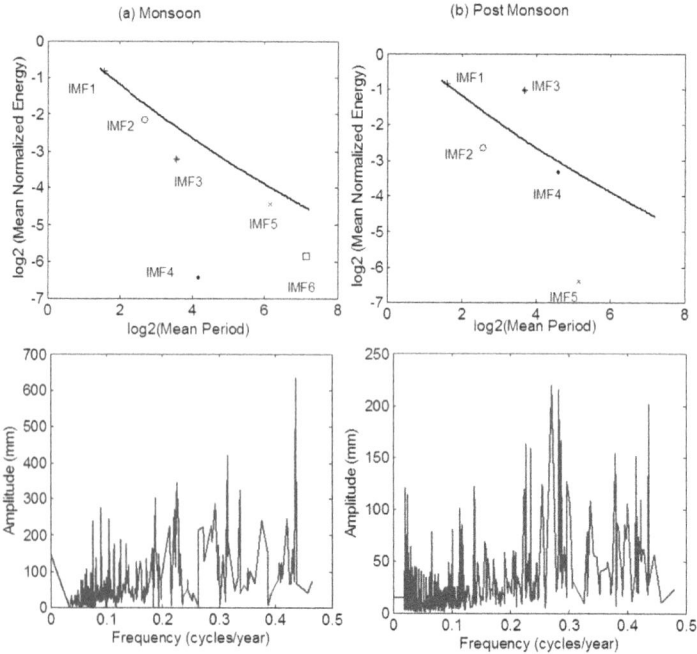

FIGURE 4.18 Statistical significance test of IMF components and the MHS obtained by integration of instantaneous frequencies of IMF components of rainfall series (a) Monsoon rainfall; (b) Post Monsoon rainfall. Upper panels give the results of statistical significance test and lower panel gives the MHS. The solid line in the upper panel represents upper confidence line at 5 % significance level.

of an oscillatory component of near decadal periodicity on the temporal variability has been reported. In the discrete wavelets, the scales which vary in dyadic powers are to be set *a priori* by the user; hence the study found that periodicity of 8 (or 16) are closest to the decadal scale. But such dyadic scales may not be in line with the process scales inherent in the data series. The CEEMDAN based decomposition is purely data adaptive and hence it could detect the near decadal scale. Such near decadal scales can also be closely associated with climatic indices such as sunspot number (of 11 years). In general, from the spectral analysis of rainfall time series data, it is noted that (i) the lower modes showed intermittency in frequency which clearly depicts the nonlinearity of the all the time series; highest variability (frequency modulation) is noticed for IMF1, which is an indicative of nonstationarity of the processes.; (ii) the highest amplitude corresponding to each IMF (concentration of dark/black color in the plot of instantaneous trajectories) is very much localized in time. But in some cases, such regions exist for longer time spells; (iii) The MHS of Kerala is characterized by multiple prominent peaks for both the time series, and extreme events can be expected during monsoon season with shorter periodicity.

4.5 INVESTIGATING MULTISCALE TELECONNECTIONS OF ISMR

The hydroclimatic teleconnection refers to the association of hydrologic variables with large scale atmospheric/oceanic oscillations from different parts of the world. To investigate hydro-climatic teleconnections, the existing practice is to perform multiscale decomposition of the time series pair and compare the periodicity of their oscillatory modes (Iyengar and Raghu Kanth 2005; Massei et al. 2007, 2011; Massei and Fournier 2012). An HHT based approach namely an innovative strategy in which the running correlation of oscillatory modes of the time series TDIC is really helpful to for teleconnection studies (Huang and Schmitt 2014), which can be easily extended to investigate the climatic teleconnections of ISMR.

4.5.1 Multiscale Investigation of Hydroclimatic Teleconnection Using HHT

In the multiscale investigation of hydroclimatic teleconnection, first a correlation analysis between the oscillatory components of hydrologic variables and different climate indices are made. Subsequently, the comparison of their residues is made in the time domain to examine the link between them in a nonstationarity perspective. The different steps involved in the procedure are presented below.

1. Decompose the time series of hydrologic variable and climate indices using the EMD method or its variants. If multiple variables are involved in the teleconnection study, MEMD can be used for decomposition of the time series signals.
2. Perform a correlation analysis by finding the correlation coefficient between the IMFs of hydrologic time series with that of a particular climatic index

time series to draw proper inferences regarding the association, in terms of periodicity.

3. Rescale the residue of hydrologic time series about its mean.
4. Compare the zero-mean residue of climatic index time series with the zero-mean residue of hydrological time series.
5. If the zero crossing of both series is nearly at same time instant, vital conclusions can be drawn regarding the association of the two signals in terms of nonstationarity.
6. Perform TDIC analysis between the respective IMFs (of comparable periodicity) to draw useful inferences on the association between the two in different time scales.

The application of the above methodology for investigating the hydroclimatic teleconnections of AISMR is demonstrated in the following sections.

4.5.2 DATA DETAILS

To explore the hydroclimatic teleconnections of ISMR, data of five different climate indices QBO, ENSO, SN, AMO, and EQUINOO were collected and used in this study. The monthly data of QBO was obtained from the website of National Oceanic and Atmospheric Administration (NOAA) Earth System Research Laboratory (www.esrl.noaa.gov/psd/data/correlation/qbo.data) for the period 1950–2012. The intensities of ElNiño events are generally assessed on the basis of the average Sea Surface Temperature (SST) over different Niño regions in the Pacific Ocean within specific latitudes and longitudes, and it was noticed that summer monsoon rainfall over India is best correlated with temperature anomaly from Niño3.4 region, which overlaps between Niño 3 and Niño 4 (Maity and Nagesh Kumar 2008, 2009). The SST data corresponding to the Niño3.4 region (120°W–170°W, 5°S–5°N) called as Oceanic Niño Index (ONI), was obtained from NOAA National Weather Service Climate Prediction Center (www.cpc.ncep.noaa.gov/data/indices/) for the period 1950–2012 and used as the ENSO index. The data of sunspot number was obtained from the solar physics group at NASA's Marshall Space Flight Centre (http://solarscience.msfc.nasa.gov/greenwch/spot_num.txt), for the same period. The relationship of AMO with monsoon rainfall was also investigated in few studies (Goswami et al. 2006; Lu et al. 2006; Zhang and Delworth 2006; Feng and Hu 2008). The data of monthly AMO indices was obtained from NOAA National Weather Service Climate Prediction Center (www.esrl.noaa.gov/psd/data/timeseries/AMO/) for the period 1950–2012. The relation between the EQUINOO and Indian monsoon was studied by few researchers (for example, Gadgil et al. 2004; Maity and Nagesh Kumar 2006a, 2006b; 2007; Maity et al. 2007; Kashid and Maity 2012). The data for zonal wind component for the region 60–90°E, 2.5°S–2.5°N) was obtained from the National Centre for Environmental prediction (NCEP) (www.esrl.noaa.gov/psd/data/gridded/data.ncep.reanalysis.html) for the same period. The negative of the anomaly of the zonal component of surface wind in the equatorial Indian ocean region (60–90°E, 2.5°S–2.5°N) is considered as EQUINOO index (Maity and Nagesh Kumar 2006a). It is to be noted that the data of climatic variables from 1950 to 2012 was used for the present analysis, as the data for most of the climate indices of 1950 onwards were available.

4.5.3 MULTISCALE DECOMPOSITION AND TELECONNECTIONS OF AISMR TIME SERIES

As a first step, the multiscale decomposition of AI monthly monsoon rainfall time series and the five climate indices for the period 1950–2012 is performed using the CEEMDAN algorithm after setting the noise parameter as 0.2 for rainfall, 0.02 for climate indices and number of realizations as 500 for all cases, following the past studies (Antico et al. 2014). The orthogonal modes obtained by decomposition of rainfall series and the different climatic indicators are presented in Figure 4.19 and Figure 4.20, respectively. The mean period of the modes obtained from the decompositions of different climate indices and AISMR are presented in Table 4.9.

Table 4.9 shows that different climate indices possess nearly same period of evolution as that of AISMR particularly for lower modes. For higher modes, the periodicities do not match, and this may be because only limited number of cycles is present in modes at larger scales.

The cross correlation analysis between the modes of AISMR time series with that of different climate indices has been performed, and the results are presented in Table 4.10.

From Table 4.10 it is noticed that there exists a very high correlation only between the trend components of the monsoon rainfall time series and different climate indices. Also few higher order modes show a reasonable correlation (highlighted in bold numbers in Table 4.10; for example, IMF6 of monsoon rainfall and IMF5/IMF6 of SN time series; IMF6 of monsoon rainfall with IMF6 of QBO, etc.). This is an important observation as it can be concluded that the relationship between climate indices and monsoon rainfall show better agreement in the low frequency part of their spectra. In order to understand the pattern of evolution in the time domain (i.e., how the changes occur over and below the mean value over the time domain) the plots of

---

TABLE 4.9

Mean period of modes of AI monthly monsoon rainfall and different climate indices for 1950–2012

Mode Number	Rainfall	Climate indices				
	AI	QBO	ENSO	SN	AMO	EQUINOO
IMF1	3.231	3.150	3.600	3.111	3.360	3.150
IMF2	6.300	9.333	8.400	6.146	7.636	6.811
IMF3	13.263	18.000	15.750	11.455	14.000	14.000
IMF4	28.000	31.500	28.000	42.000	28.000	25.200
IMF5	42.000	63.000	50.400	126.000	84.000	50.400
IMF6	84.000	126.00	126.00	126.00	126.00	126.000
Residue	252.000	252.00	252.00	252.00	252.00	252.000

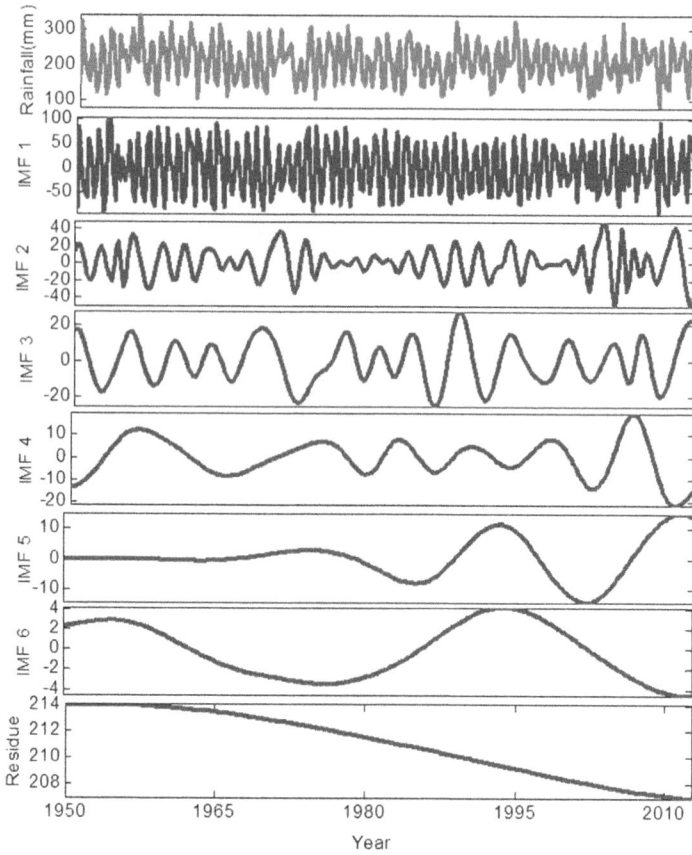

FIGURE 4.19   Decomposition of monthly monsoon rainfall series (1950–2012).

trend components of monsoon rainfall with the trend component of all five climate indices are made and the results are presented in Figure 4.21. A comparison of the trends of different climate indices with that of rainfall shows that the different indices show very good agreement with rainfall. Also, the changes over the mean are found to be occurring in similar manner. Further it is noticed that the zero crossing occurs more or less at the same instant of time in all cases. This also establishes a strong long-term association of the different climate indices with Indian monsoon rainfall.

It is to be noted that in the present correlation analysis, the correlation between IMFs are computed by considering the entire time span. But there is a possibility that the IMFs of climate indices and rainfall may show strong positive (negative) correlation for shorter time spells. In addition, at some of the time spells, such correlation can be negative where some other spells it may be positive. To illustrate this aspect, the IMFs of ENSO and AISMR are plotted and presented in Figure 4.22.

From Figure 4.22, it can be seen that overall the correlation between the two are negative in most of the high frequency modes and the residue, but the two are positively correlated for the shorter time spells ~1960–1965 in IMF2, ~1978–1982

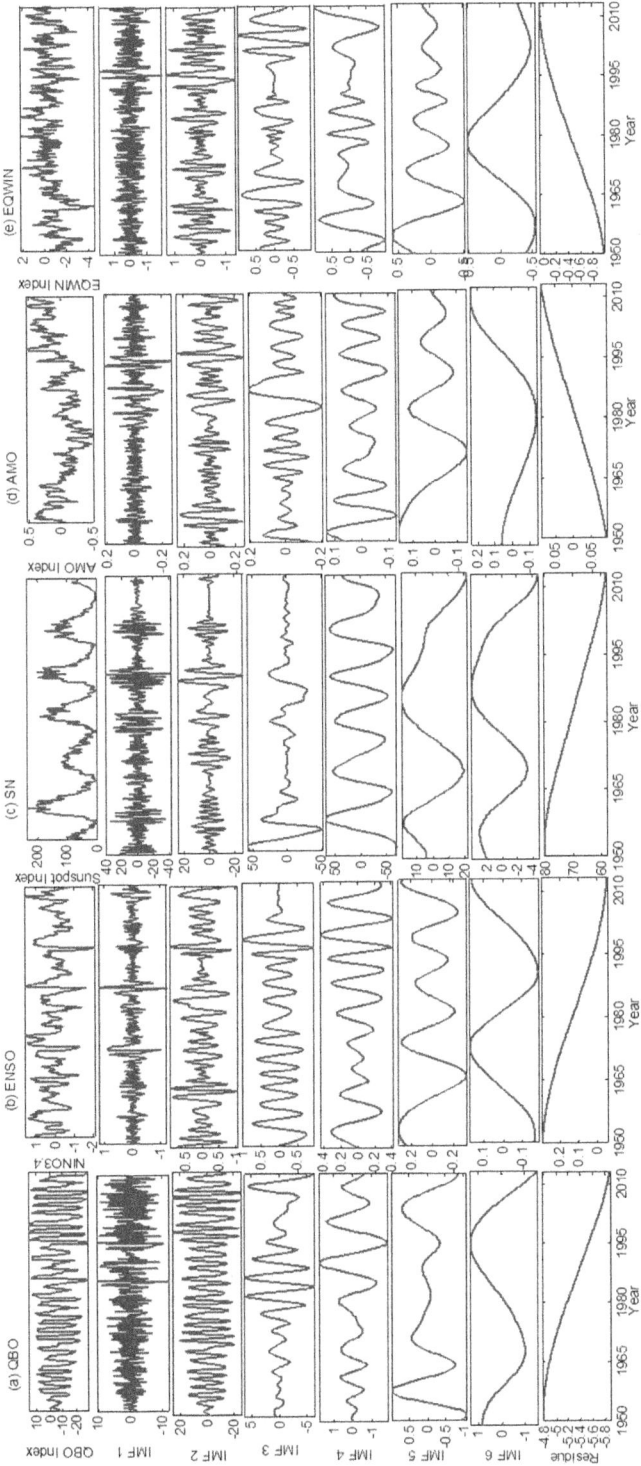

FIGURE 4.20 Multiscale decomposition of five climatic indicators using CEEMDAN. (From Adarsh 2017).

TABLE 4.10
Correlation coefficients between modes of monthly monsoon rainfall (1950–2012) of AI and IMF components of different climate indices

Mode of Monsoon Rainfall	Climate Index	Mode of climate index						
		IMF1	IMF2	IMF3	IMF4	IMF5	IMF6	Residue
	QBO							
IMF1		−0.085	0.044	−0.03	−0.012	−0.013	−0.018	−0.002
IMF2		0.015	0.019	0.078	0.015	0.039	0.018	0.027
IMF3		0.019	−0.007	*−0.407*	0.12	−0.048	−0.061	−0.024
IMF4		0.016	0.042	−0.133	−0.115	*0.563*	*0.217*	0.156
IMF5		−0.009	0.01	0.026	−0.084	*−0.395*	*−0.264*	−0.111
IMF6		−0.031	−0.016	0.049	−0.023	−0.027	*0.879*	0.150
Residue		0.010	0.035	0.096	0.035	−0.064	−0.181	***0.999***
	ENSO							
IMF1		−0.031	−0.022	0.022	0.000	−0.038	−0.014	0.000
IMF2		−0.114	−0.347	−0.165	−0.018	0.038	−0.008	−0.025
IMF3		0.065	−0.084	−0.399	−0.183	0.085	−0.013	0.015
IMF4		−0.025	0.035	−0.104	−0.100	−0.055	0.170	−0.134
IMF5		0.016	−0.005	0.020	0.094	−0.154	−0.188	0.088
IMF6		0.035	−0.032	0.038	−0.110	−0.202	***0.945***	−0.142
Residue		−0.003	−0.014	0.078	−0.097	−0.147	0.119	***−0.992***
	SN							
IMF1		−0.010	−0.018	−0.041	−0.017	0.014	−0.002	−0.001
IMF2		0.018	−0.014	−0.032	−0.053	−0.017	−0.007	0.025
IMF3		−0.009	0.013	0.068	0.157	−0.042	−0.042	−0.019
IMF4		−0.017	0.039	−0.102	0.073	*0.329*	*0.273*	0.151
IMF5		0.030	−0.028	0.129	−0.171	*−0.354*	*−0.328*	−0.104
IMF6		0.015	0.003	0.117	−0.079	***0.653***	***0.810***	0.173
Residue		−0.008	0.009	0.127	−0.058	*0.238*	0.175	***0.996***
	AMO							
IMF1		0.085	0.020	−0.023	0.010	0.032	0.010	0.001
IMF2		−0.065	−0.044	0.017	−0.045	−0.015	−0.006	−0.027
IMF3		−0.063	−0.072	0.074	0.124	0.054	0.054	0.018
IMF4		0.024	0.154	0.059	*−0.206*	−0.003	−0.173	−0.142
IMF5		−0.007	−0.030	0.122	−0.014	*−0.401*	0.129	0.102
IMF6		−0.032	0.024	0.136	0.007	*0.418*	−0.026	−0.181
Residue		0.022	0.077	−0.027	0.023	*0.320*	***−0.559***	***−0.992***
	EQUINOO							
IMF1		*−0.349*	−0.068	−0.013	−0.017	−0.028	0.000	0.000
IMF2		−0.036	−0.154	−0.032	−0.023	0.010	−0.003	−0.024
IMF3		−0.006	−0.094	−0.151	0.011	−0.069	−0.037	0.011

TABLE 4.10 (Continued)
Correlation coefficients between modes of monthly monsoon rainfall (1950–2012) of AI and IMF components of different climate indices

Mode of Monsoon Rainfall	Climate Index	Mode of climate index						
		IMF1	IMF2	IMF3	IMF4	IMF5	IMF6	Residue
IMF4		0.012	0.051	–0.166	*–0.256*	0.082	0.020	–0.126
IMF5		–0.035	–0.071	0.021	0.034	–0.041	0.004	0.079
IMF6		–0.047	0.026	–0.037	–0.132	*0.208*	*–0.496*	–0.157
Residue		0.027	0.028	0.023	–0.098	*0.207*	0.141	***–0.985***

Note: Italic numbers indicates significant correlation (p value <0.05) and bold letters indicates moderate to strong correlation (say >±0.5).

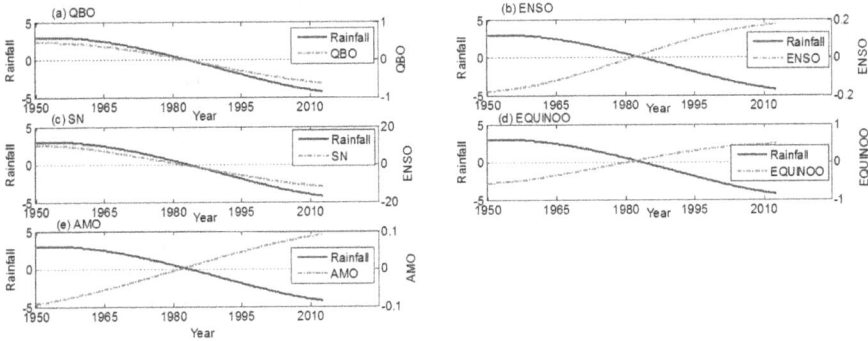

FIGURE 4.21 Trend components of climate indices and monsoon rainfall of AI during 1950–2012.

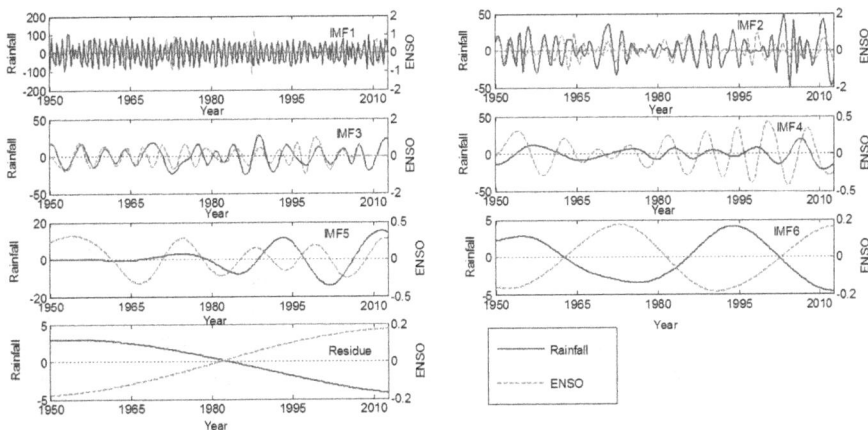

FIGURE 4.22 Plots of orthogonal modes of ENSO and AISMR time series. (From Adarsh 2017).

in IMF2 and IMF3, ~1994–1996 in IMF3, etc., i.e., the correlation coefficient of two data series on the whole domain alone may not reveal the possible relationship between them particularly when the processes are intermittent or contain drift and trends or they are nonstationary. In addition, it is noticed that the relation between IMF4 of ENSO and rainfall is very weak for most of the time spells. A strong positive correlation is observed between ENSO and AISMR in IMF6 for most of the time and in IMF5 for the period ~1987–2007. The relation in IMF5 was noticed with a phase shift for the period ~1997–2007.

Hence it can be concluded that for the complex relation between rainfall and climate indices, certain processes involved may correlate with each other in one scale but not in others. Also, the correlation between rainfall and climate indices might have changed from strong positive correlation to strong negative correlation in some of the scales (i.e., for IMFs). This evidence is clear on considering the much-investigated relation between ENSO and ISMR by the comparison of their IMFs. The two events are negatively correlated for the residue component (correlation coefficient of -0.992), but a strong positive correlation for the IMF6 (correlation coefficient of 0.945) (Table 4.10). This is noteworthy because negative correlation at one scale would counteract positive correlation at another scale, resulting low overall correlation between rainfall and climate indices. In this case the overall correlation between ENSO and monsoon rainfall series is only -0.21, while the correlation values between rainfall series and QBO, SN, AMO and EQUINOO are 0.01, -0.001, 0.013 and -0.33, respectively. It is to be noted that the above correlation values are quite low and the correlation coefficient depicts only the linear association between the different climatic variables with ISMR, while the true association between them might be of nonlinear in characteristics. But the climate forcing of lower periodicity may last for shorter time spells and they might have significant impact on the hydrological processes of the regime. To identify such local associations, correlation analysis considering shorter time spells can be adopted, and an in-depth intrinsic correlation analysis of IMFs may give more insight to understand the linkages between climate indices and Indian monsoon rainfall. Therefore, a time dependent intrinsic correlation analysis of AISMR time series is performed.

4.5.4   TDIC ANALYSIS OF AISMR TIME SERIES

TDIC is calculated among the different pairs of IMFs (of climate index and monsoon rainfall) and TDIC plots are prepared for the first four IMFs, as the periodicities of monsoon rainfall and different climate indices matches reasonably well only for the first four IMFs. The TDIC analysis of higher order modes (IMF5 onwards) and of all climate indices does not pass the student t-test in most of the time scales for most of the time spells. In the implementation of TDIC analysis, first the instantaneous frequencies (and hence periods) are to be estimated, for which the NHT-DQ scheme is used. Using the instantaneous periods, the TDIC algorithm is invoked and the TDIC plots are prepared for different cases. To illustrate the applicability of TDIC analysis, first the relationship between ENSO and rainfall is considered and the TDIC plots for different IMFs are presented in Figure 4.23.

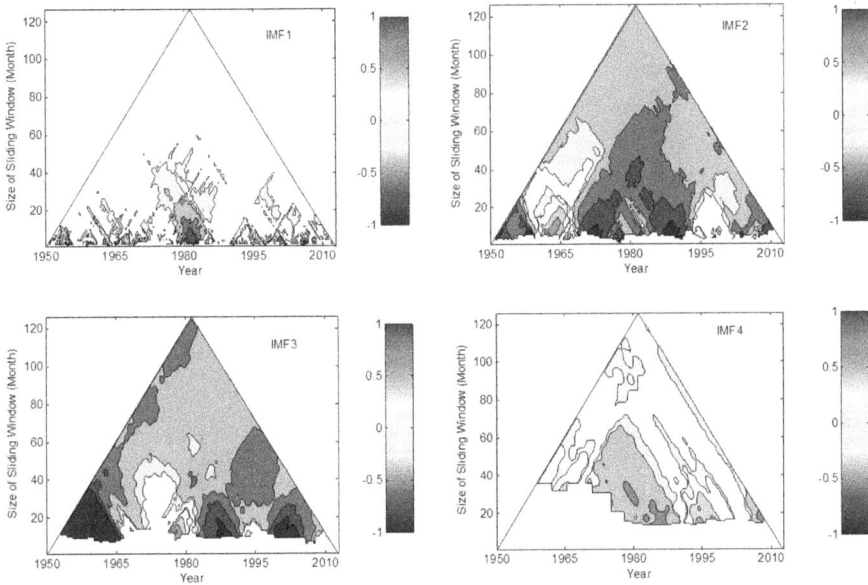

FIGURE 4.23 TDIC plots of ENSO and monsoon rainfall. The white space of the TDIC plot means that the correlation between ENSO and monsoon rainfall is not significant at 5% significance level.

The bottom contour of the triangular plots depicts the instantaneous frequency and hence a shift of the plots to larger time scales can be noticed in higher order IMFs (of low frequency modes). The white spaces in the plots represents that such correlations are not statistically significant. It is to be noted that in the teleconnection studies, the appended monthly data sets of monsoon period from different years is considered, while HHT is one such tool recommended even for the spectral analysis of datasets with irregular periodicity (Huang et al. 2009a; Cong and Chetouani 2009; Rahman et al. 2015). This subsection aims to examine the association between the variables in multiple time scales whatever be the scale (periodicity) associated with them. However, it is important to perform the appropriate time scale conversion while periodicity becomes the central focus of discussion. ENSO shows a long-range negative correlation with AISMR for IMF2 (~8.4*4/12 = 3-year periodicity) and IMF3 (~15.7*4/12 = 5-year periodicity). So it can be inferred that both modes may be contributed by same physical processes like westerly wind bursts or oscillatory patterns such as Madden Julian Oscillations (MJO) that may vary in intra seasonal scales (Wang and Picaut 2004). However localized positive correlation of ENSO with AISMR is observed in different modes in different shorter time spells. For example, IMF2 in ~1964–1966, IMF3 in ~1979–1982, IMF4 in 1995–1998 etc. It is well known that the strongest ElNiño of the century (1997–1998) triggered an Indian Ocean Dipole (IOD), which resulted in above-average rainfall during the period 1997–1998 (Kumar et al. 2006). From the TDIC analysis (Figure 4.23) it is found that for IMF4, the correlation between the AISMR and ENSO time series

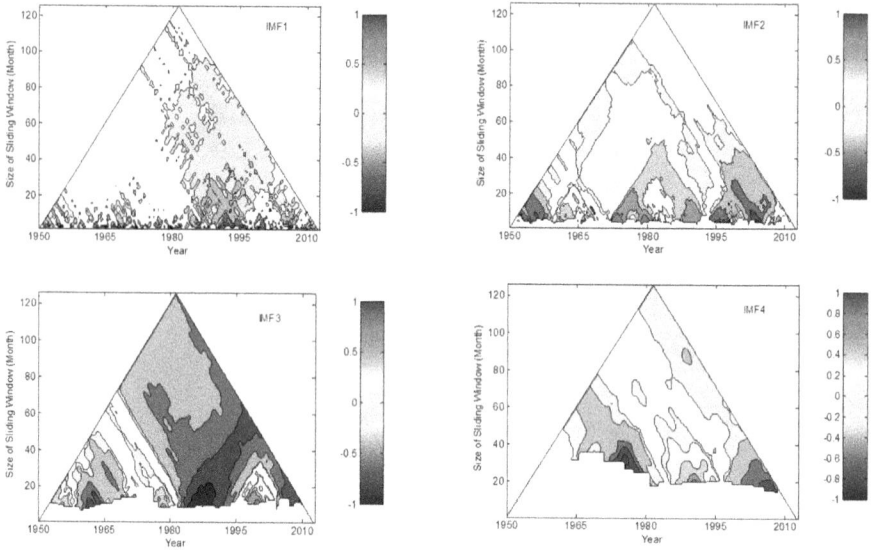

FIGURE 4.24 TDIC plots of QBO and monsoon rainfall. The white space of the TDIC plot means that the correlation between QBO and monsoon rainfall is not significant at 5% significance level.

is strongly positive during 1995–1998. This matches with the results presented by Chen et al. (2010) who established the relation between IOD and ENSO using the TDIC analysis. Moreover, it is observed that there are frequent alterations in nature of correlation between ENSO and AISMR in the high frequency mode IMF1. Such dynamics (transition from positive correlation to negative and vice versa) is more apparent for smaller window size (i.e., in the lower part of the TDIC plot). Also, it is well understood that in 2001–2002, Indian monsoon weakened due to the effect of ENSO, and it is observed that IMF2 and IMF3 shows very strong negative correlation with rainfall. The TDIC plots of QBO, SN, EQUINOO and AMO links are presented in Figures 4.24–4.27

Figure 4.24 shows that IMF1 of QBO is negatively correlated with that of AISMR in almost all scales. A direct correlation between QBO and ISMR was observed in the period ~1967–1994 for the IMF2; but in the recent past (~1995–2010), the IMF2 also shows an anticorrelating behavior. For IMF3, the relation between QBO and AISMR is primarily a long-range negative. But some significant direct relation between the two events was noticed during short-term spells 1953–1955, 1978–1980 and 1997–1998. From Figure 4.25, it is noticed that the high frequency modes of SN (IMF1 and IMF2) show statistical significance only upto window size of ~60 months. Overall, the correlation between the SN and AISMR is weak and shows rich dynamics (i.e., frequent alterations in correlations) in pattern in different TDIC plots of SN time series.

From Figure 4.27 a weak positive correlation is observed between AMO and AISMR in different modes. Interestingly, all the first four modes show a strong direct

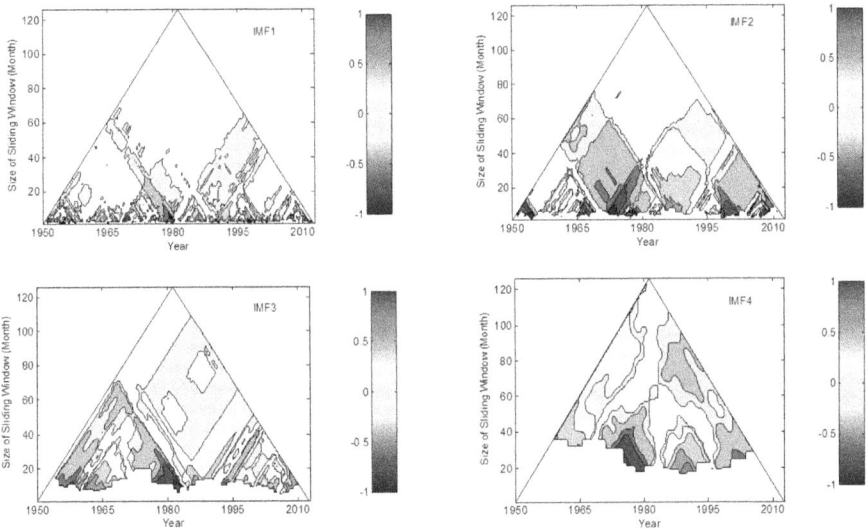

FIGURE 4.25 TDIC plots of SN and monsoon rainfall. The white space of the TDIC plot means that the correlation between SN and monsoon rainfall is not significant at 5% significance level.

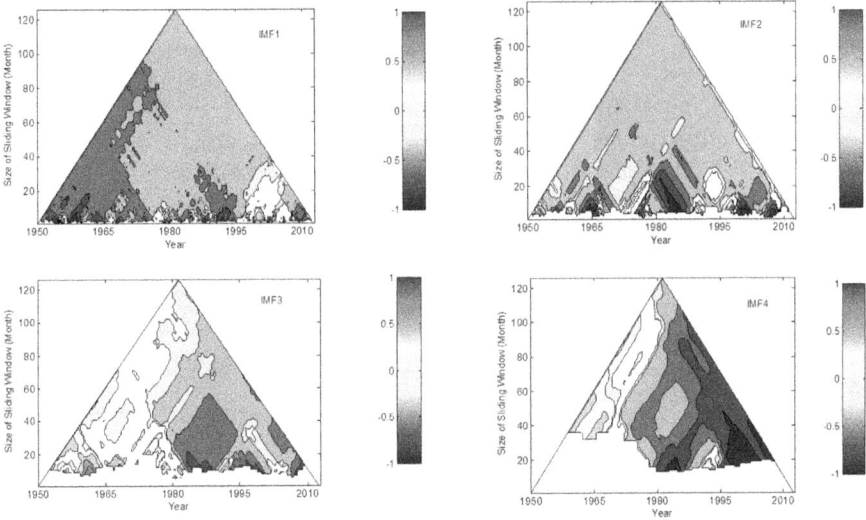

FIGURE 4.26 TDIC plots of EQUINOO and monsoon rainfall. The white space of the TDIC plot means that the correlation between EQUINOO and monsoon rainfall is not significant at 5% significance level.

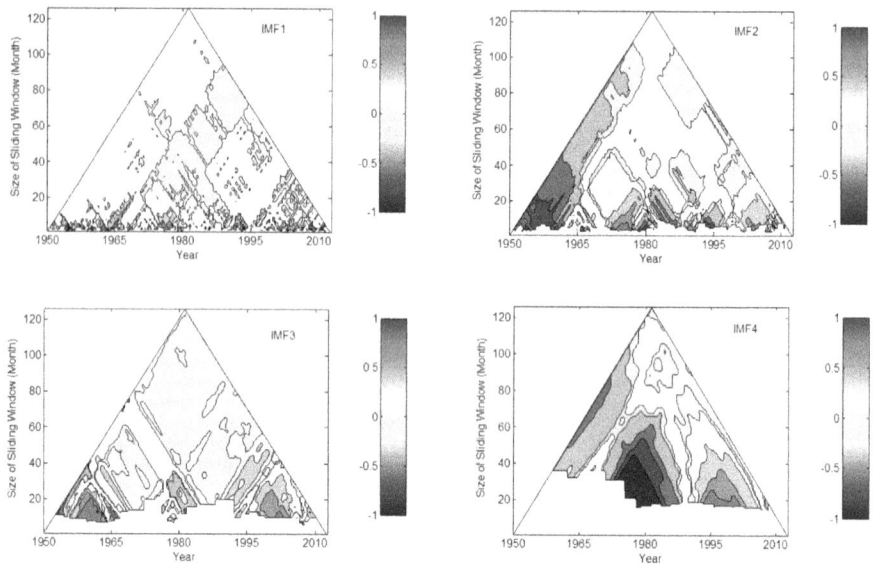

FIGURE 4.27   TDIC plots of AMO and monsoon rainfall. The white space of the TDIC plot means that the correlation between AMO and monsoon rainfall is not significant at 5% significance level.

correlation with monsoon rainfall in the recent past (~1995–2010). More specifically, it is observed that during 1997–1998 there exists a strong correlation between the two in all modes, which supports the earlier findings of the modulation of Indian monsoon by AMO and its link with ENSO (Goswami et al. 2006; Dong et al. 2006). From Figure 4.26, it is noticed that most of the modes of EQUINOO show an anti-correlation with the monsoon rainfall, both at shorter and longer scales, but very localized direct correlations are observed between the two at lower time scales of less than 2 years (for example, IMF3 in 1950s, IMF1 in 2000s, etc.). The significant negative correlation of IMF1 of EQUINOO and AISMR is quite different from that of other indices considered in this study. Also at a particular time spell, the relation of different climate indices with monsoon rainfall may be quite different, for example, during 1950–1960 (for scale ranges < 20 months) the correlation of modes of QBO, EQUINOO etc. with that of monsoon rainfall is dominantly negative, while it is positive for different modes of AMO.

4.5.5   MEMD-BASED TDIC FOR HYDROCLIMATIC TELECONNECTION

To examine the efficacy of the proposed MEMD-TDIC approach for teleconnection studies, the most debated ENSO-monsoon rainfall link in Kerala subdivision is considered (Krishnakumar et al. 2009; Gadgil et al. 2004; Kumar et al. 2006). First, the TDIC plots for first five IMFs obtained by the individual decomposition (by EMD) of ENSO and rainfall are presented in Figure 4.28.

For implementation of MEMD, the rainfall data and the data of for climatic oscillations QBO, ENSO, EQUINOO and AMO together constitute the multivariate

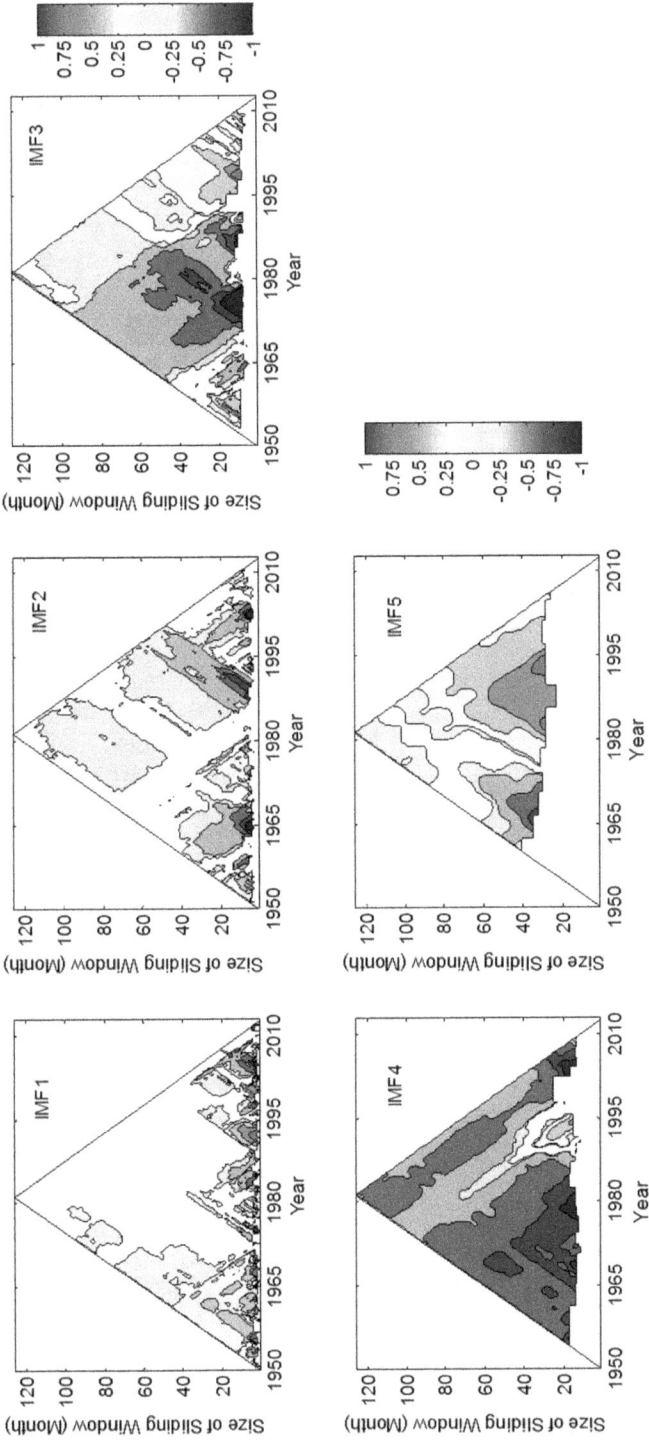

FIGURE 4.28 TDIC plots between the IMFs of ENSO and summer monsoon rainfall in Kerala obtained by using EMD algorithm. The void spaces of the plot depicts that correlation coefficients are not significant at 5% significance level. (From Adarsh 2017).

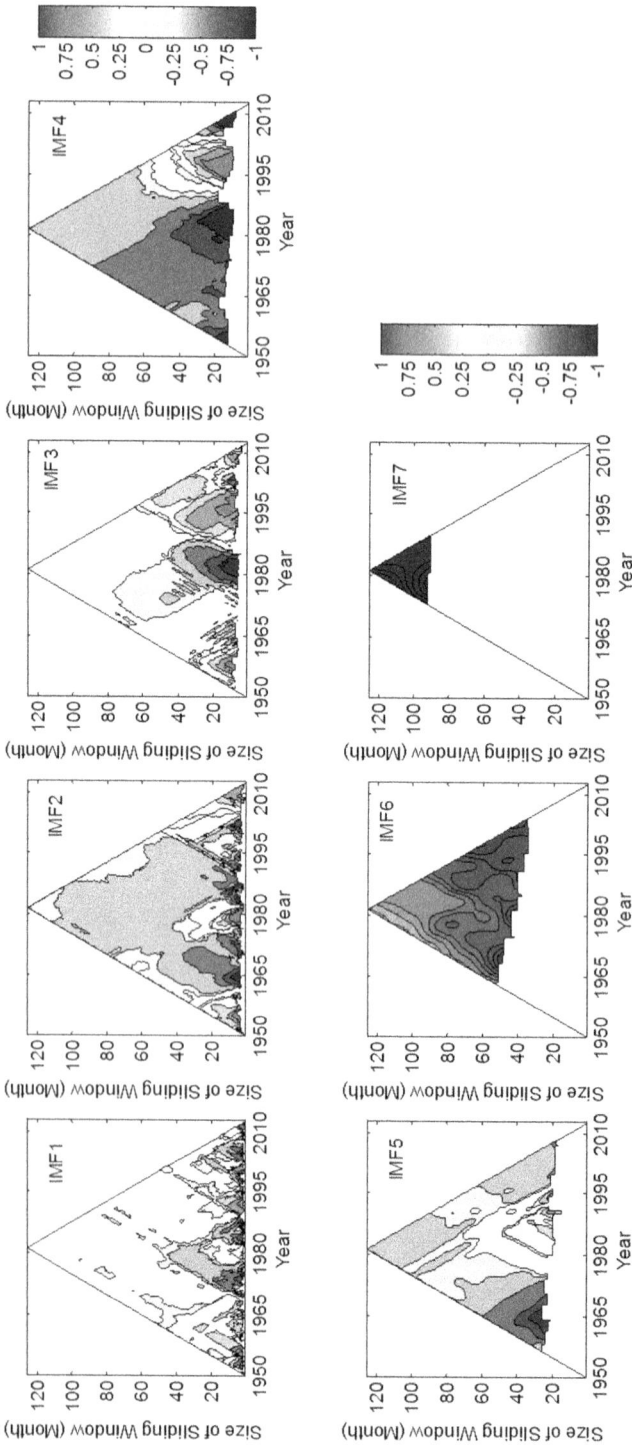

FIGURE 4.29 TDIC plots between IMFs of ENSO and summer monsoon rainfall in Kerala obtained by using MEMD algorithm. The void spaces of the plot depicts that correlation coefficients are not significant at 5% significance level. (From Adarsh 2017).

dataset. To apply MEMD methodology, the number of directions is selected as 64, and the upper and lower bound of threshold parameters are chosen as 0.075, 0.75 and the splitting parameter is chosen as 0.075, following the suggestions made in the past works (Hu and Si 2013; Rilling et al. 2003), and performed the decomposition of multivariate dataset. MEMD resulted in seven IMFs and residue for all the time series. Now by employing the MEMD algorithm, the TDIC analysis is performed and the results are presented in Figure 4.29.

In the ENSO-rainfall link (Figure 4.29), a dominancy of negative association is noticed with the exception of positive associations in inter-decadal scales of 4 years and 7 years (IMF3 and IMF4) during the 1995–2000s. A careful perusal of Figure 4.28 and Figure 4.29 show that in Figure 4.29, a strong positive association is displayed after 1995 for both IMF3 and IMF4, while it is completely lacking in IMF4 and rather weak in IMF3 of Figure 4.28. It may be noted that the process scales of these modes (4–7 years) fairly matches with the periodicity of a typical ElNiño event and hence the display of correlations in these modes needs special attention. Also it is noted that during 2002–2003, a strong negative correlation between the two is displayed in different modes in Figure 4.17, while it is not observed in Figure 4.29. Hence it is inferred that such true associations are captured in a much better way by the proposed MEMD-TDIC approach for teleconnection studies, when compared with EMD based TDIC approach.

4.6 DEVELOPING HOURLY IDF CURVES BASED ON MULTIVARIATE EMPIRICAL MODE DECOMPOSITION AND SCALING THEORY

Intensity Duration Frequency (IDF) curve is one of the important hydrologic relation needed for urban infrastructure design. The IDF curves of short duration (such as hourly durations) are quite useful for design of urban storm water drains, while the rainfall observations are available in coarse resolution such as daily time step. The information on the time scale invariance property of rainfall is a useful mean for esti-mation of IDF relationships for shorter durations from the data in coarse time reso-lution through the disaggregation operation in time domain. The traditional methods for the preparation of IDF curve may fail in such circumstances and the scale invari-ance theory can be used to solve this problem. Also, the output of downscaling studies of rainfall are primarily available in monthly time resolutions and in such cases this concept can be used in the derivation of IDFs under a changing climate scenario (Ali and Mishra 2014; Afrin et al. 2015; Herath et al. 2015; Chandrarupa et al. 2015). In the past, different researchers made attempts to derive shorter duration rainfall IDFs from coarse resolution data based on the scaling property (Nhat et al. 2008; Bara et al. 2009). Stemming from the scale invariance theory, alternative approaches were also evolved to construct rainfall IDF relationships, in which the recent method proposed by Kuo et al. (2013) involves the computation of scaling exponents based on EEMD. As the EEMD method perform the decomposition adaptively (considering the complexity of the data), the number of modes obtained by the decomposition of different intensity series may differ. This may introduce errors in the estimation of scaling exponents. Hence it is believed that the decomposition by the identification of common time scales in different intensity series is a better alternative and the common scales present in different rainfall intensity series can be captured by the Multivariate EMD (MEMD) method, in which the decomposition of all of the rain-fall intensity series can be performed in single step. This simplifies the procedure considerably and helps to maintain equal number of modes for different time series. Therefore, this study proposes an alternative method for developing the rainfall IDF curves of shorter duration from coarse resolution data employing the scaling theory, in which the representative scaling exponent is computed based on MEMD.

4.6.1   MEMD-EV-PWM Framework for Developing Hourly IDF Curves from Coarse Resolution Rainfall Data

For deriving the IDF relationships based on the scaling property, first the MEMD method is used for the decomposition of rainfall intensity time series of different durations simultaneously. The logarithmic plot between the Probability Weighted Moments (PWM) of the orthogonal modes obtained from MEMD and the duration gives the scaling exponent, and finally the IDF relationships are derived based on Extreme Value (EV) formulations involving scaling exponents. Hence the method can be designated as MEMD-EV-PWM method, as described in Chapter 2. First, this method is applied for rainfall data from Colaba station, India, and the results are compared with that obtained by the frequency factor method. Then the method is applied for the derivation of IDF relationships for different hourly durations from the daily rainfall data of Thiruvananthapuram city in the state of Kerala and the results are compared with that obtained by an existing empirical formula. The encour-aging results motivated to apply the proposed method to develop the hourly IDF relationships for seven more major cities in Kerala from the daily rainfall data.

4.6.2   Developing IDF Curves for Colaba, Mumbai, from Hourly Data

The city of Mumbai (18.50°N, 72.52°E), located on the western coast of India is popularly known as economic capital of India. The population of the city is growing very fast and the city surrounded by sea, hills and creeks and therefore has limitations on horizontal growth (to provide required infrastructure facilities) due to scarcity of land. The existing drainage system of Mumbai city has a network of roadside surface drains, an underground drainage system in the island city, major and minor nallah dis-charging storm water, either into four main rivers or directly into the creek and finally to the Arabian sea (Zope et al. 2016). The hourly data for the period 1969–2004 are used for preparation of IDF curves by using the EV-PWM approach based on scaling theory and MEMD. First, the annual maximum rainfall intensity series for different durations (d = 1, 2, 6, 12, 24hrs) is prepared. For the decomposition, the threshold parameters of MEMD are selected as 0.075, 0.75 and 0.075, following the guidelines in literature (Hu and Si 2013) to avoid the chances of any over decomposition. The decomposition resulted in a total of five modes ranging from high to low frequency and the results of decomposition are presented in Figure 4.30. The PWM values are computed for each mode and the log-linear fitting of PWM and durations are made, for different moment orders and the results of fitting are presented in Table 4.11. The

TABLE 4.11
Scale exponent (β) and R² of orthogonal modes of original intensity series from Colaba station for different moment orders

Mode	Scale exponent (β) and R² statistics	Moment order			
		1	2	3	4
IMF1	β	−0.4879	−0.4646	−0.4522	−0.4415
	R²	0.9349	0.9483	0.9483	0.9445
IMF2	β	−0.4248	−0.4332	−0.4351	−0.4351
	R²	0.9117	0.9247	0.9279	0.9280
IMF3	β	−0.4865	−0.5228	−0.5360	−0.5440
	R²	0.8914	0.9022	0.9051	0.9066
IMF4	*β*	*−0.3330*	*−0.3213*	*−0.3168*	*−0.3132*
	R²	*0.5788*	*0.5749*	*0.5716*	*0.5664*
Residue	β	−0.4912	−0.4584	−0.4311	−0.4059
	R²	0.9918	0.9888	0.9799	0.9653

Note: The numbers in italic indicate that the IMF component is discarded while forming the new intensity series.

results indicate that the R² statistics of the fitting of IMF4 is the least. Therefore, this mode is excluded and a new time series is formed by considering all the remaining modes. The decomposition of new series also resulted in six modes and the fitting statistics gave R² values of all modes consistently greater than 0.85. The mean representative scaling exponent (β) is found to be −0.47397. This exponent is used in the following scaling to obtain the IDF for different durations and return period:

$$d = \frac{\mu + \sigma\left(-\ln\left(-\ln\left(1-\frac{1}{T}\right)\right)\right)}{d^{-\beta}} \quad (4.2)$$

where μ and σ are the location parameter and scale parameter respectively given by $\mu = \mu_D(D)^{-\beta}$ and $\sigma = \sigma_D(D)^{-\beta}$; i_d is the rainfall intensity for duration d, T is the return period of i_d.

To validate the results obtained by the proposed method, the IDF curves are also prepared by the classical frequency factor method. The IDF obtained by both methods are presented in Figure 4.31. To enable a comparison of the IDF developed by the proposed method, the average root mean square deviation of the intensity values computed for each curve is estimated. The values are 2.2, 5.7, 8.1, 13.3 and 15.5 for different return periods 2 years, 5 years, 10 years, 50 years and 100 years, respectively. Also, the maximum rainfall of annual series of different durations is noted and the corresponding intensity values are estimated and presented in Figure 4.31. It is noted that for two high intense short duration rainfalls, the estimates by the

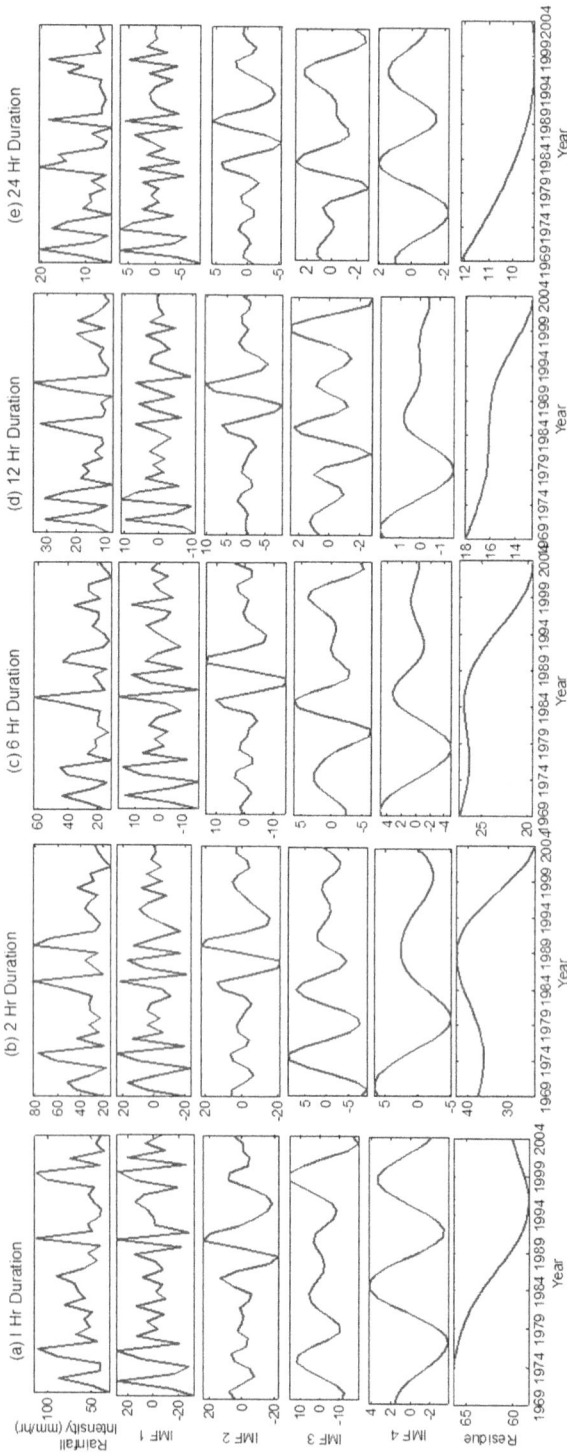

FIGURE 4.30   Decomposition of rainfall intensity time series of different durations for Colaba station rainfall (1969–2004).

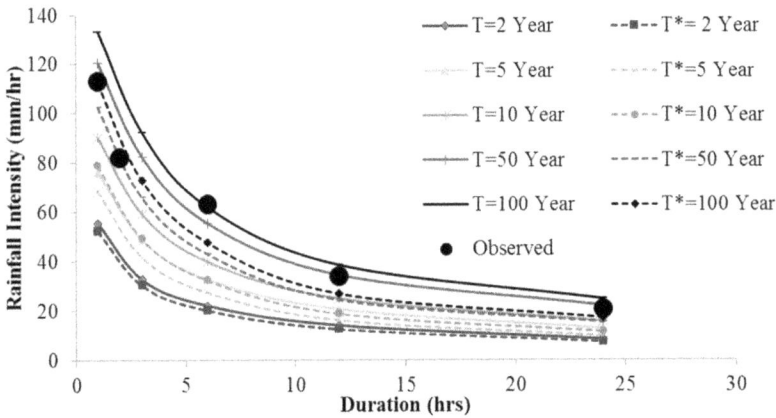

FIGURE 4.31 Comparison between IDF curves for Colaba station by MEMD-EV-PWM method and frequency factor method. T is the return period for frequency factor method and T* is the return period for MEMD-EV-PWM method.

MEMD-EV-PWM are more accurate than that by frequency factor methods, which is more relevant for the urban drainage system design for the city of Mumbai. Thus, it is noticed that the proposed method is equally good in developing IDF curves as the frequency factor method and more deviation is noticed for curves of larger return periods. Thus, the comparison confirms that the proposed method is effective in developing IDF curves from hourly data.

4.6.3 DEVELOPMENT OF HOURLY IDF CURVES FOR CITIES IN KERALA FROM DAILY DATA

The proposed approach for developing subdaily IDF curves is further applied to daily rainfall data of eight cities in Kerala. First, the hourly IDF curves of the city of Trivandrum (Thiruvananthapuram) is prepared form the daily time series data obtained for a period of 25 years (1991–2015). The annual maximum rainfall intensity for different durations (-day maximum, 2-day maximum, 6-day maximum, 12-day maximum and 24-day maximum etc.) is estimated and subsequently, the proposed procedure is implemented. The decomposition resulted in five modes. A plot between log(PWM) and log(duration) is made for the different modes, for different moment orders ($q = 1–4$), and the plot is presented in Figure 4.32. The slope of the log(PWM) *versus* log(duration) gives the scale exponent (β). From Figure 4.32 it is noted that, for IMF3, the fitting statistics (R^2 and β) for different moment orders are not consistent therefore IMF3 is discarded to form the new intensity series (IMF124R).

On proceeding with four more trials, it is noticed that the slope becomes constant and the average scale exponent is found to be −0.6133. This exponent is used in the scaling IDF formula (Equation 4.2) to derive the subdaily IDFs. Then, the empirical equation proposed by Rambabu et al. (1979) in the following form is used to prepare the IDF curve of Trivandrum station.

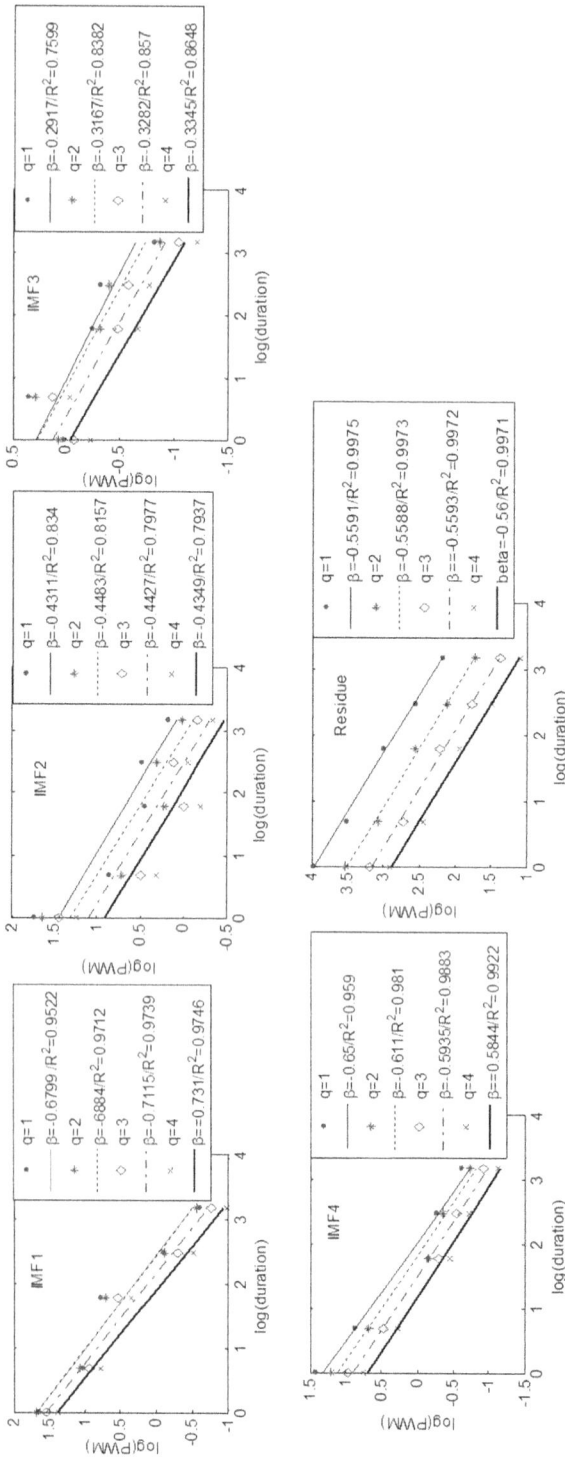

FIGURE 4.32 Plot between log(PWM) and log(duration) for different modes of daily data from Thiruvananthapuram station, for different moment orders (q = 1 to 4).

$$I = \frac{KT^a}{(t+b)^n} \tag{4.3}$$

where I is the rainfall intensity in cm/hr, T is the return period in years, t is the storm duration in hours, K, a, b and n are constants developed for different stations and zones of India. For Trivandrum station, the constants K, a, b and n are 6.762, 0.1536, 0.5, 0.8158, respectively. The IDF curves prepared by MEMD-EV-PWM method and that prepared by the empirical formula for Trivandrum city data are presented in Figure 4.33. The comparison between the two IDF curves is made by estimating average root mean square deviation of curves for different return period. Here these values are 6.1, 5.4, 5.9, 6.8 and 6.3, respectively, for curves of T = 2, 5, 10, 50 and 100 years. Hence for this case, the deviations are fairly consistent and show no definite pattern with return period.

Some studies reported that it is preferable to perform the rainfall projection studies under a changing climate scenario at monthly time scale and the scaling theory need to be applied for developing the IDF curves of different regions based on future rainfall (Afrin et al. 2015; Herath et al. 2015; Chandrarupa et al. 2015; Shreshta et al. 2017). The methodology being general, the MEMD-EV-PWM method can be used to develop hourly IDF curves from monthly data also. To enable a comparison between the IDF curves derived from two coarse resolution time scales (daily and monthly), subdaily IDFs for the city of Thiruvananthapuram is prepared also based on monthly data of the period 1989–2012 following the presented MEMD-EV-PWM approach (for which the representative scaling exponent was −0.6217). It is to be noted that in the scaling IDF formula (Equation 4.2), the scaling factor to be considered as 720 in deriving the hourly IDF curve from monthly time series. The plot is given in Figure 4.34, which matches fairly well with those derived from daily data.

In the next exercise, the daily data of six cities in Kerala (Kollam, Alappuzha, Kottayam, Eranakulam, Kozhikode and Kannur) for the period 1991–2015 and for the

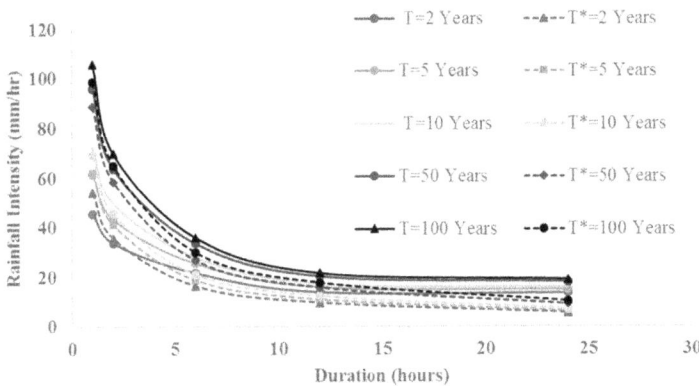

FIGURE 4.33 Comparison between hourly IDF curves prepared for Thiruvananthapuram station by MEMD-EV-PWM method and empirical method. T is the return period for empirical method and T* is the return period for MEMD-EV-PWM method. (From Adarsh 2017).

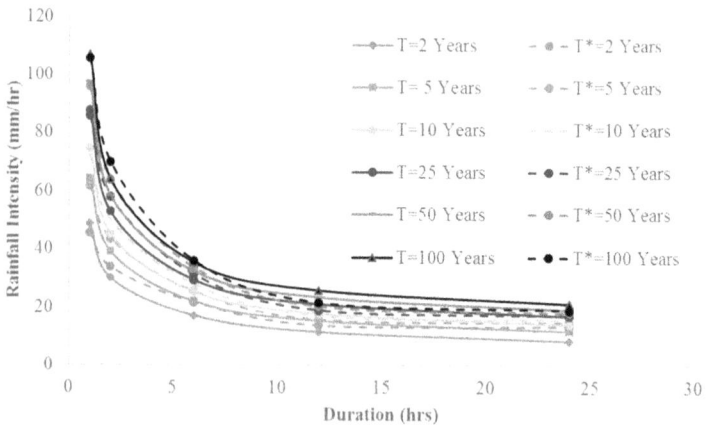

FIGURE 4.34 Comparison between IDF curves for Thiruvananthapuram city derived from daily data and monthly data using the MEMD-EV-PWM method. T is the return period for daily data and T* is the return period for monthly data. (From Adarsh [2017]).

period 1991–2014 for the city of Thrissur are considered to develop the IDF relations for these cities. During the implementation of the MEMD-EV-PWM method, in most of the cases five IMFs are resulted and three to five cycles were sufficient to obtain the representative scale exponent. The distribution fitting parameters (μ and σ)of intensity series of different durations and the representative scaling exponents obtained from the daily rainfall series of all eight cities in Kerala are summarized in Table 4.12 These parameters will be helpful in developing IDF curves for different return period (T) using the scaling IDF formula (Equation 4.1). For example, the IDF curves of 1hr duration rainfall takes the general form: $i_{d,T} = \dfrac{148.6 + 47.3F^{-1}(1 - 1/T)}{d^{-0.6133}}$.

The resulting IDF curves are presented in Figure 4.35.

The above applications demonstrate that the proposed MEMD-EV-PWM approach is effective in deriving IDF relationships for shorter durations (like subdaily) from data of larger temporal resolutions (daily/monthly). The proposed method identifies common scales present within different intensity series, which makes the uniqueness in number of decomposition modes, which finally makes the procedure simpler when compared with the decomposition based on EMD/EEMD.

4.7 CLOSURE

This chapter presented some of the important applications in the field of hydrology using rainfall data of different temporal resolutions ranging from daily to annual. The applications include the extraction of nonlinear trend, spectral characterization, teleconnection and scaling analysis of rainfall. The extraction of nonlinear trend will be more appealing in capturing the effect of climatic changes as the residue component can effectively portray the trend. The EMD graphical approach in conjunction with nonparametric statistical tests is found to be very much efficient in deciphering

TABLE 4.12
Distribution fitting parameters and mean scale exponents for daily rainfall data from different stations in Kerala

City	d = 1 hr		d = 2 hr		d = 6 hr		d = 12 hr		d = 24 hr		Scale Exponent
	μ	σ	μ	σ	μ	σ	μ	σ	μ	σ	
Thiruvananthapuram	148.60	47.26	92.40	27.53	53.21	14.68	35.31	11.52	24.69	10.43	−0.6133
Kollam	150.70	34.19	102.48	19.93	55.82	12.79	38.01	8.77	26.25	5.91	−0.5869
Alappuzha	137.78	36.02	102.85	27.67	60.04	13.90	42.14	9.88	31.76	7.42	−0.5836
Kottayam	151.47	41.44	104.51	23.61	64.87	18.68	46.39	14.68	34.74	9.98	−0.5873
Eranakulam	164.96	39.32	114.83	30.27	70.74	25.64	49.43	15.96	36.77	10.10	−0.5548
Thrissur	194.08	108.11	127.79	47.29	71.67	18.07	50.43	10.84	38.85	7.83	−0.5836
Kozhikode	190.21	59.74	140.82	46.67	80.41	17.18	58.21	13.77	45.41	13.47	−0.5229
Kannur	224.52	82.06	160.03	45.20	96.84	22.71	67.13	14.55	49.99	12.06	−0.6057

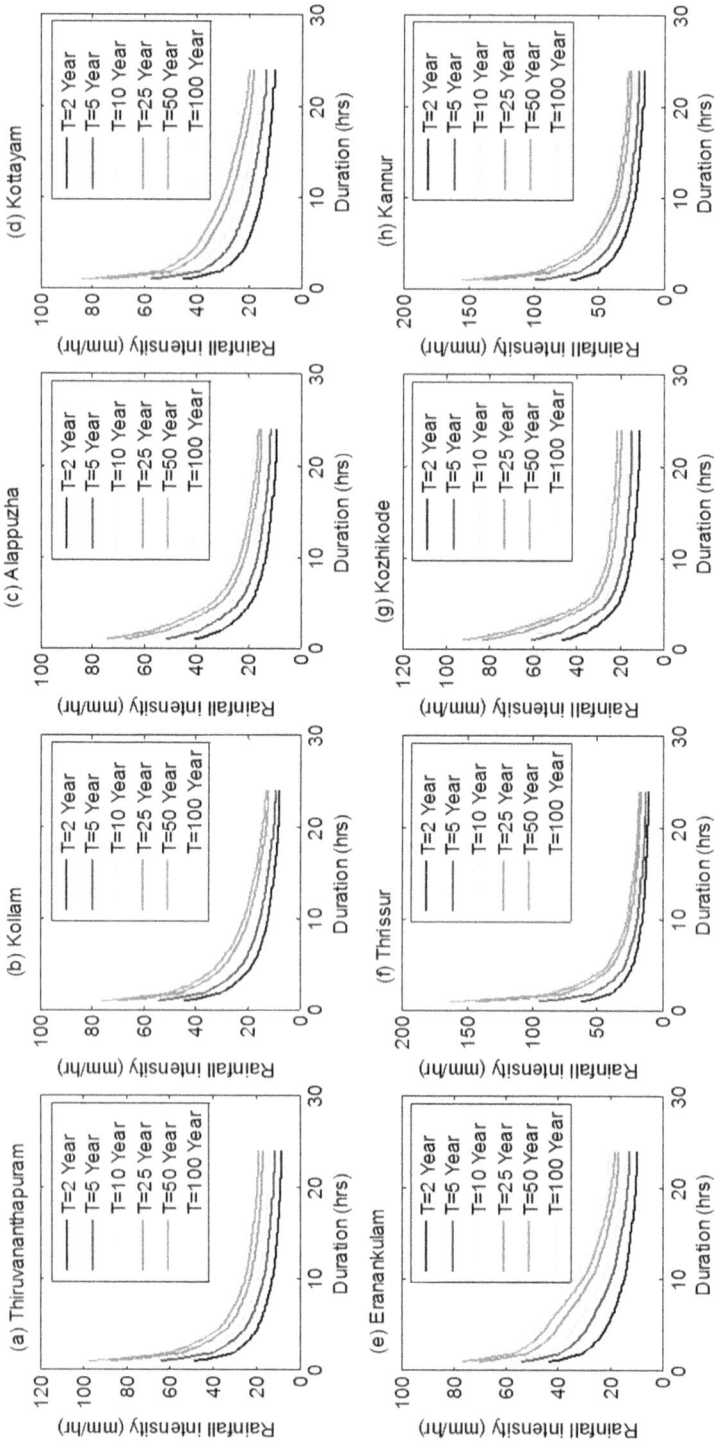

FIGURE 4.35   Hourly IDF curves for the eight cities in Kerala prepared by MEMD-EV-PWM hybrid method. (From Adarsh 2017).

the trend of different seasonal and annual rainfall datasets at smaller (subdivisonal) as well as larger (macro-regions) spatial regions in India. Even though climatic teleconnections of Indian summer monsoon rainfall are well debated, exploring it in a multiscale perspective was quite a rare practice. The TDIC extension of HHT is proven to be a potential candidate for performing such analysis on considering the case of AISMR dataset. The nature and strength of associations between climatic oscillations and rainfall display changes with respect to the time scale and over the time domain. The TDIC method could capture the well debated above average rainfall of 1997 and drought of 2002. Even though the development of hydrological tool like IDF curve has much practical significance, the scarcity of hourly data is a major challenge in the Indian scenario. Therefore, the rainfall disaggregation is very significant in dealing with the development of IDF curves and subsequent design of urban drainage systems. The MEMD-PWM coupling with the scaling theory is found to be a promising method for rainfall disaggregation and circumventing such challenges on finer time scale datasets in India. Thus, HHT is proven to be a potential tool for many practical hydrological problems including trend analysis, teleconnection studies and developing IDF curves through rainfall disaggregation.

REFERENCES

Adarsh, S. 2017. *Multiscale characterization of hydrologic time series using mathematical transforms*. Ph.D. thesis, IIT Bombay, India.

Adarsh, S., Janga Reddy, M. 2015. Trend analysis of rainfall in four meteorological subdivisions of southern India using non-parametric methods and wavelet analysis. International Journal of Climatology **35**(6): 1107–1124.

Afrin, S., Islam, M.M., Rehman, M.M. 2015. Development of IDF curve for Dhaka city based on scaling theory under future precipitation variability due to Climate Change. *International Journal of Environmental Research and Development* **6**(5): 332–335.

Ali, H., Mishra, V. 2014. Development of sub-daily Intensity Duration Frequency (IDF) curves for major urban areas in India. (Paper ID Number: 14778), AGU Fall meeting 2014, San Francisco, USA.

Ananthakrishnan, R., Parthasarthy, B., Pathan, J.M. 1979. Meteorology of Kerala. *Contributions to Marine Sciences* **60**: 123–125.

Antico, A., Schlotthauer, G., Torres, M.E. 2014. Analysis of hydro-climatic variability and trends using a novel empirical mode decomposition: Application to Parana river basin. *Journal of Geophysical Research: Atmospheres* **119**(3): 1219–1233.

Bara, M., Kohnova, S., Gaal, L., Szolgay, J., Hlavcov, K. 2009. Estimation of IDF curves of extreme rainfall by simple scaling in Slovakia. *Contributions in Geophysics and Geodesy* **39**(3): 187–206.

Barnhart, B.L. 2011. *The Hilbert Huang transform theory and development*. Ph.D thesis, University of Iowa, Lincoln NE.

Barnhart, B.L., Eichinger, W.E. 2011a. Analyzing sunspot variability using empirical mode decomposition. *Solar Physics* **269**(2): 439–449.

Barnhart, B.L., Eichinger, W.E. 2011b. Empirical mode decomposition applied to solar irradiance, global temperature, sunspot number and CO_2 concentration data. *Journal of Atmospheric and Solar-Terrestrial Physics* **73**(2011): 1771–1779.

Campbell, W.H., Blechman, J.B., Bryson, R.A. 1983. Long period tidal forcing of Indian monsoon rainfall: A hypothesis. *Journal of Climate and Applied Meteorology* **22**: 287–296.

Chattopadhyay, J., Bhatla, R. 2002. Possible influence of QBO on teleconnections relating Indian summer monsoon rainfall and sea-surface temperature anomalies across the equatorial pacific. *International Journal of Climatology* **22**(1): 121–127.

Chandrarupa, R., Sha, U., Mujumdar, P.P. (2015). Model and parameter uncertainty in IDF relationships under climate change. *Advances in Water Resources* **79** (2015): 127–139.

Chen, X., Wu, Z., Huang, N.E. 2010. The time-dependent intrinsic correlation based on the empirical mode decomposition. *Advances in Adaptive Data Analysis* **2**: 233–265.

Claud, C., Pascal, T. 2007. Revisiting the possible links between the quasi-biennial oscillation and the Indian summer monsoon using NCEP R-2 and CMAP fields. *Journal of Climate* **20**: 773–787.

Cong, Z., Chetouani, M. 2009. Hilbert-Huang transform based physiological signals analysis for emotion recognition. *International Symposium on in Signal Processing and Information Technology* (ISSPIT) Ajman, UAE, pp. 334–339.

Dong, B., Sutton, R.T., Scaife, A.A. 2006. Multidecadal modulation of El Niño–Southern Oscillation (ENSO) variance by Atlantic Ocean sea surface temperatures. *Geophysical Research Letters* 33 L08705, doi:10.1029/2006GL025766.

Feng, S., Hu, Q. 2008. How the North Atlantic multidecadal oscillation may have influenced the Indian summer monsoon during the past two millennia. *Geophysical Research Letters* **35**, L01707, doi:10.1029/2007GL032484.

Franceschini, S., Tsai, C.W. 2010. Application of Hilbert-Huang transform method for analyzing toxic concentrations in the Niagara River. *Journal of Hydrologic Engineering* **15**(2): 90–96.

Gadgil, S., Vinayachandran, P.N., Francis, P.A., Gadgil, S. 2004. Extremes of the Indian summer monsoon rainfall, ENSO and equatorial Indian Ocean oscillation. *Geophysical Research Letters* 31 L12213 doi:10.1029/2004GL019733.

Goswami, B.N., Madhusoodanan, M.S., Neema, C.P., Sengupta, D. 2006. A physical mechanism for North Atlantic SST influence on the Indian summer monsoon. *Geophysical Research Letters* **33**, L02706, doi:10.1029/2005GL024803.

Guhathakurta, P., Rajeevan, M. 2008. Trends in the rainfall pattern over India. *International Journal of Climatology* **28** (11): 1453–1469.

Hamed, K.H., Rao, A.R. 1998. A modified Mann-Kendall trend test for autocorrelated data. *Journal of Hydrology* **204**: 182–196.

Herath, H.M.S.M., Sarukkalige, P.R., Nguyen, V.TV. 2015. Downscaling approach to develop future sub-daily IDF relations for Canberra airport region, Australia. *Proceedings of IAHS* **369**: 147–155.

Hu, W., Si, B-C. 2013. Soil water prediction based on its scale-specific control using multivariate empirical mode decomposition. *Geoderma* **193–194**: 180–188.

Huang N.E., Shen Z., Long S.R., Wu M.C., Shih H.H., Zheng Q., Yen N.C., Tung C.C., Liu H.H. 1998. The empirical mode decomposition and the Hilbert spectrum for non-linear and non-stationary time series analysis. *Proceedings of Royal Society London, Series A* **454**: 903–995.

Huang Y., Schmitt F.G. 2014. Time dependent intrinsic correlation analysis of temperature and dissolved oxygen time series using empirical mode decomposition. *Journal of Marine Systems* **130**: 90–100.

Huang, Y., Schmitt, F.G, Lu, Z., Liu, Y. 2009a. Analysis of daily river flow fluctuations using empirical mode decomposition and arbitrary order Hilbert spectral analysis. *Journal of Hydrology* **373**: 103–111.

Huang, N.E, Wu, Z., Long, S.R., Arnold, K.C., Blank, K., Liu, T.W. 2009b. On instantaneous frequency. *Advances in Adaptive Data Analysis* **1**(2): 177–229.

Iyengar, R.N., Raghu Kanth, T.S.G. 2005. Intrinsic mode functions and a strategy for forecasting Indian monsoon rainfall. *Meteorology and Atmospheric Physics* **90**: 17–36.

Kashid, S.S., Maity, R. 2012. Prediction of monthly rainfall on homogeneous monsoon regions of India based on large scale circulation patterns using genetic programming. *Journal of Hydrology* **454**: 26–41.

Kendall, M.G. 1975. *Rank Correlation Methods*. Charles Griffin: London.

Kripalani, R.H., Kulkarni, A. 1997a. Climatic impacts of El Niño/La Nina on the Indian monsoon: A new perspective. *Weather* **52**: 39–46.

Kripalani, R.H., Kulkarni A. 1997b. Rainfall variability over south East Asia- Connections with Indian monsoon and ENSO extremes: New perspectives. *International Journal of Climatology* **17**: 1155–1168.

Krishnakumar, K.N., Rao, G.S.L.H.V.P., Gopakumar, C.S. 2009. Rainfall trends in twentieth century over Kerala, India. Atmospheric Environment **43**: 1940–1944.

Kumar, K.K., Rajagopalan, B., Hoerling, M., Bates, G., Cane, M. 2006. Unraveling the mystery of Indian monsoon failure during El Niño. *Science* **314**: 115–119.

Kumar, K.N., Rajeevan, M., Pai, D.S., Srivastava, A.K., Preethi, B. 2013. On the observed variability of monsoon droughts over India. *Weather and Climate Extremes* 1: 42–50.

Kumar, V., Jain, S.K., Singh, Y. 2010. Analysis of long-term rainfall trends in India. *Hydrological Sciences Journal* **55**(4): 484–496. Kuo, C., Gan, T., Chan, S. 2013. Regional Intensity-Duration-Frequency curves derived from ensemble empirical mode decomposition and scaling property. *Journal of Hydrologic Engineering* **18**(1): 66–74.

Lu, R., Dong, B., Ding, H. 2006. Impact of the Atlantic Multidecadal Oscillation on the Asian summer monsoon. *Geophysical Research Letters* 33, L24701, doi:10.1029/2006GL027655.

Maity, R., Nagesh Kumar, D. 2006a. Bayesian dynamic modeling for monthly Indian summer monsoon rainfall using El Niño-Southern Oscillation (ENSO) and Equatorial Indian Ocean Oscillation (EQUINOO). *Journal of Geophysical Research - Atmospheres* 111, D07104, doi:10.1029/2005JD006539.

Maity, R., NageshKumar, D. 2006b. Hydroclimatic association of monthly summer monsoon rainfall over India with large-scale atmospheric circulation from tropical Pacific Ocean and Indian Ocean region. *Atmospheric Science Letters* 7(4): 101–107.

Maity, R., Nagesh Kumar, D. 2007. Hydroclimatic teleconnection between global sea surface temperature and rainfall over India at subdivisional monthly scale. *Hydrological Processes* **21**(14): 1802–1813.

Maity, R., Nagesh Kumar, D. 2008. Basin-scale streamflow forecasting using the information of large-scale atmospheric circulation phenomena. *Hydrological Processes* 22(5): 643–650.

Maity, R., Nagesh Kumar, D. 2009. Hydroclimatic influence of large-scale circulation on the variability of reservoir inflow. *Hydrological Processes* **23**(6): 934–942.

Maity, R., Nagesh Kumar, D., Nanjundiah, R.S. 2007. Review of hydroclimatic teleconnection between hydrologic variables and large-scale atmospheric circulation patterns with Indian perspective. *ISH Journal of Hydraulic Engineering* **13**(1): 77–92.

Mann, H.B. 1945. Non-parametric tests against trend. *Econometrica* **13**: 245–259.

Massei, N., Durand, A., Deloffre, J., Dupont, J.P., Valdes, D., Laignel, B. 2007. Investigating possible links between the North Atlantic oscillation and rainfall variability in northwestern France over the past 35 years. *Journal of Geophysical Research* 112, D09121, doi:10.1029/2005JD007000.

Massei, N., Fournier, M. 2012. Assessing the expression of large scale climatic fluctuations in the hydrologic variability of daily Seine river flow (France) between 1950–2008 using Hilbert-Huang Transform. *Journal of Hydrology* 448–449: 119–128.

Massei, N., Laignel, B., Rosero, E., Motelay-Massei, A., Deloffre, J., Yang, Z.L., Rossi, A. 2011. A wavelet approach to the short-term to pluridecennal variability of streamflow in the Mississippi river basin from 1934 to 1998. *International Journal of Climatology* **31**: 31–43.Narasimha, R., Kailas, S.V. 2001. A wavelet map of monsoon variability. *Proceedings of Indian National Science Academy* **67**(3): 327–341.

Nhat, L.M., Tachikawa, Y., Sayama, T., Takara, K. 2008. Estimation of sub-hourly and hourly IDF curves using scaling properties of rainfall at gauged site in Asian Pacific region. *Annuals of Disaster Prevention Research Institute*, Kyoto University, No. 51 B.

Papoulis A. 1986. Probability, Random Variable and Stochastic Processes, 2nd ed., McGraw-Hill, New York.

Pokhrel, S., Chaudhari, H.S., Saha, S.K., Dhakate, A., Yadav, R.K., Salunke, K., Mahapatra, S., Rao, S.A. 2012. ENSO, IOD and Indian summer monsoon in NCEP climate forecast system. *Climate Dynamics* **39**: 2143–2165.

Rahman, Md. A., Chetty, M., Bulach, D., Wangikar, PP. 2015. Frequency decomposition based gene clustering. *Neural Information Processing-Lecture Notes in Computer Science*, pp. 170–181.

Rajeevan, M., Pai, D.S., Dikshit, S.K., Kelkar, R.R. 2004. IMD's new operational models for long-range forecast of southwest monsoon. *Current Science* **86**(3): 422–431.

Rambabu, Tejwani, K.K, Agrawal, M.C., Bhusan, L.S. 1979. *Rainfall intensity duration-return period equations & nomographs of India*, CSWCRTI, ICAR, Dehradun, India.

Rilling G., Flandrin P., Goncalves P. (2003). On empirical mode decomposition and its algorithms, In *Proceedings of IEEE-EURASIP Workshop on Nonlinear Signal and Image Processing NSIP-03*, Grado (Italy), pp. 8–11.

Sang Y.F., Wang Z., Liu C. 2014. Comparison of the MK Test and EMD method for trend identification in hydrological time series. *Journal of Hydrology* **510**: 293–298.

Sen, P.K. 1968. Estimates of the regression co-efficient based on Kendall's tau. *Journal of the American Statistical Association* **63**: 1379–1389.

Shreshta, A., Babel, M.S., Weesakul, S., Vojinovic, Z. 2017. Developing Intensity–Duration–Frequency (IDF) curves under climate change uncertainty: The case of Bangkok, 4 Thailand. *Water* **9** (145), doi:10.3390/w9020145.

Torres, M.E., Colominas, M.A., Schlotthauer, G., Flandrin, P. 2011. A complete ensemble empirical mode decomposition with adaptive noise. *IEEE International conference on Acoustic Speech and Signal Processing*, Prague 22–27 May 2011, pp. 4144–4147.

Unnikrishnan P., Jothiprakash V. 2015. Extraction of non-linear trends using singular spectrum analysis. *Journal of Hydrologic Engineering*, 10.1061/(ASCE)HE.1943-5584.0001237, 05015007.

Usoskin, I.G., Mursula, K. 2003. Long-term solar cycle evolution: review of recent developments. *Solar Physics* **218**: 319–343.

Wang, C., Picaut, J. 2004. Understanding ENSO physics - A review. In *Earth's Climate: The Ocean-Atmosphere Interactio*. Geophysical Monograph Series, Volume 147. Edited by C. Wang, S.-P. Xie, and J. A. Carton, pp. 21–48, Washington, DC: American Geophysical Union.

Wu, Z., Huang, N.E. 2004a. A study on the characteristics of white noise using the empirical mode decomposition method. *Proceedings of the Royal Society A: Mathematical, Physical and Engineering Sciences* **460**: 1597–1611.

Wu, Z., Huang, N.E. 2004b. Statistical significance test of intrinsic mode functions. In *Hilbert-Huang Transform and its Applications*. Edited by: Norden E. Huang (NASA Goddard Space Flight Center, USA), Samuel S P Shen (University of Alberta, Canada). Singapore: World Scientific Publishing. pp. 149–169.

Wu Z., Huang N.E., Long S.R., Peng C.K. 2007. On the trend, detrending and variability of nonlinear and non-stationary time series. *Proceedings of National Academy of Science USA*, **104:** 14889–14894.

Wu L., Cao C.C., Hsu T., Jao K., Wang Y. 2011. Ensemble empirical mode decomposition on storm surge separation from sea level data. *Coastal Engineering Journal* **53**(3): 223–243.

Zhang, R., Delworth, T.L. 2006. Impact of Atlantic multidecadal oscillations on India/Sahel rainfall and Atlantic hurricanes. *Geophysical Research Letters* **33**, L17712, doi:10.1029/2006GL02626.

Zope, P.E., Eldho, T.I., Jothiprakash, V. 2016. Development of rainfall intensity duration frequency curves for Mumbai city, India. *Journal of Water Resource and Protection* **8**: 756–765.

5 Multiscale Characterization of Streamflow and Sediment Load Using HHT

5.1 BACKGROUND

Characterization of streamflow and sediment load time series is of paramount importance in the planning and management of hydraulic structures. Semi arid rivers experiencing pronounced wet and dry seasons often display high variability in the sediment-streamflow relationship (Gray et al. 2014). Larger temporal scale corresponds to a slower variation of the physical quantity in time (such as the bed elevation changes) while smaller time scale corresponds to rapid variations (such as variations of flow). Identification and selection of such multiple scales associated with sediment concentration time series data from natural channels may help for accurate modeling of sediment transport processes and for development of suitable bed load transport formulas. Variability in suspended sediment concentration and streamflow is influenced by internal mechanism such as anthropogenic effects and external mechanisms such as climate oscillations. A number of studies have investigated changes in sediment load and streamflow, associated underlying causes and subsequent impacts (e.g., Zhang et al. 2008, 2009, 2012, 2013, 2014; Chen et al. 2010b) using the techniques such as scanning t-test/F-test, coherence analysis and Continuous Wavelet Transform (CWT). Earlier applications of HHT on daily streamflow and suspended sediment concentration reported high degree of intermittency in the high frequency part of their spectra, which is typical character of turbulence (Kuai and Tsai 2012). Kuai and Tsai (2012) applied the HHT for identification of varying time scale in sediment transport process by gathering both bed load and suspended sediment concentration data from Rio Grande river in New Mexico. To investigate the linkage between streamflow and sediment concentration in a multiscale framework and its teleconnections with climate forcing this study proposes HHT based TDIC approach. This typical multiscaling behavior could also be considered as character of fractals. Earlier research works proved that extending the HSA to arbitrary moment orders may lead to an efficient tool namely AOHSA for fractal (multifractal) characterization of time series (Huang et al. 2009a). However, its practical application is still scarce, and we present the application of AOHSA for the multifractal characterization of streamflow and TSS time series of Mahanadi river, India. Few of the past studies also reported that the climate oscillations can influence streamflows at large spatiotemporal resolutions (Maity and Nagesh Kumar 2008, 2009). As multiple climate variables simultaneously influence the streamflow variability, for investigating the teleconnections of streamflow, an MEMD-based TDIC approach is also proposed and demonstrated its usefulness with the datasets of monthly inflows to Hirakud dam, India.

5.2 MULTISCALE CHARACTERIZATION OF STREAMFLOW AND TSS CONCENTRATION FROM MAHANADI RIVER, INDIA

In the following section, details of the datasets used in the study are provided. Subsequently the implementation of CEEMDAN-TDIC approach for investigating the link between streamflow and Total Suspended Sediment (TSS) concentration is presented.

5.2.1 DESCRIPTION OF DATASETS

Mahanadi river is one of the important intermittent rivers in India, in which one of the major multi-purpose reservoir project Hirakud is located. The river has a length of 850 kilometers approximately rises in the highlands of Chattisgarh and flows through Chattisgarh and Orissa states in east-central India drains an area of 141,589 km² to discharge into the Bay of Bengal along the east coast of India. The major part of the upper reaches of the Mahanadi river catchment encompasses most of the areas of Chattisgarh state. The Basantpur river gauging station in Mahanadi river is operated by the Central Water Commission (CWC) of India is located a few kilometres upstream of the Hirakud dam, the most important and largest among different control structures of the basin and the upper Mahanadi cover 60% of the basin nearly 83,000 km² of basin area. Tikrapara station is located in the central Mahanadi basin (in Orissa state between 82°E and 86°E longitude and 19°N and 22°N latitude having area of nearly 43,000 km²). For better understanding of the dynamic processes of sediment transport and water discharge needs monthly data (Walling and Fang 2003; Zhang et al. 2009). Therefore, daily records of streamflow and TSS concentration at Basantpur and that at Tikrapara for the period 1973–2007 are collected from Central Water Commission (www.india-wris.nrsc.gov.in/) and are converted to monthly data for the present study. The location map showing the Basantpur and Tikrapara stations and Hirakud dam are shown in Figure 5.1.

5.2.2 MULTISCALE SPECTRAL ANALYSIS OF STREAMFLOW AND TSS CONCENTRATION SERIES

In the following, first the results of multiscale decomposition of streamflow (SF) and TSS concentration time series by employing the CEEMDAN algorithm are presented. Subsequently, the results of Hilbert spectral analysis of IMF components of both the

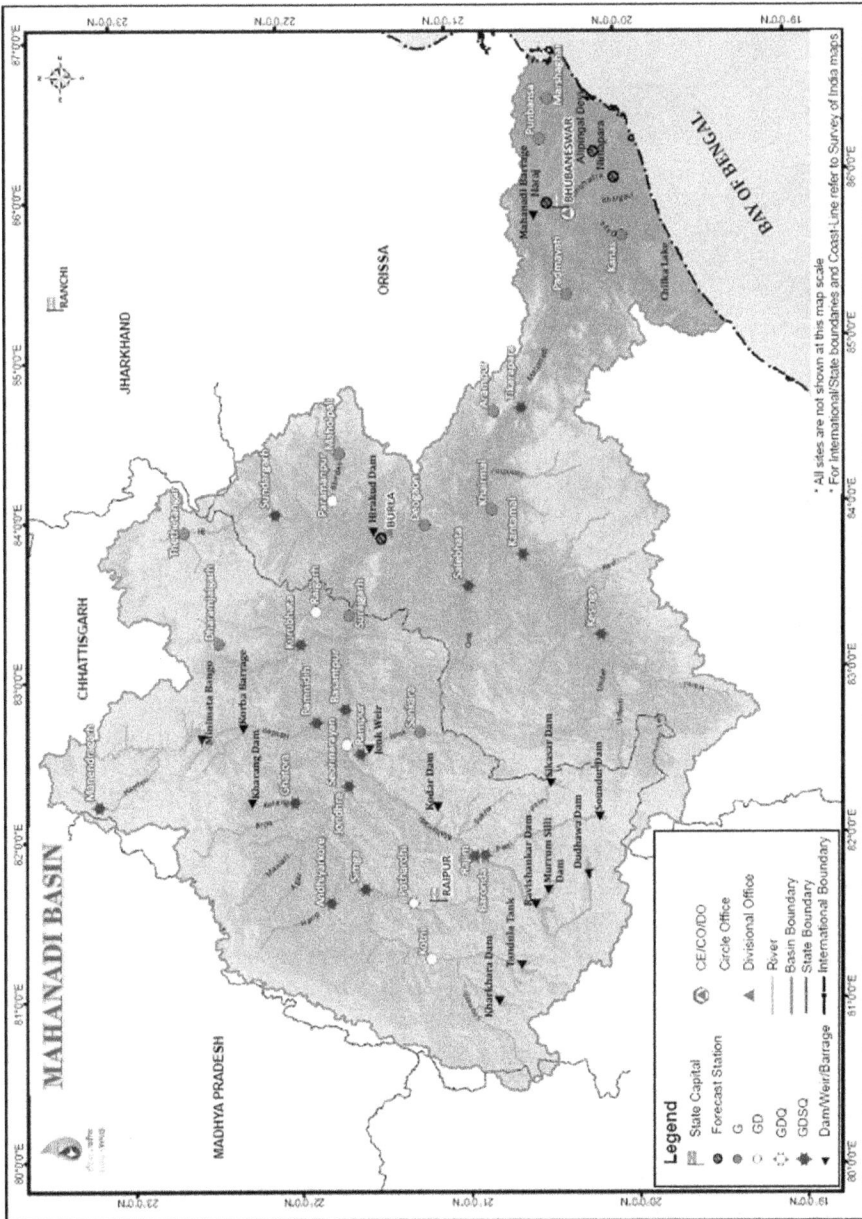

FIGURE 5.1 Map of Mahanadi basin. (From http://india-wris.nrsc.gov.in/wrpinfo/index.php?title=Mahanadi).

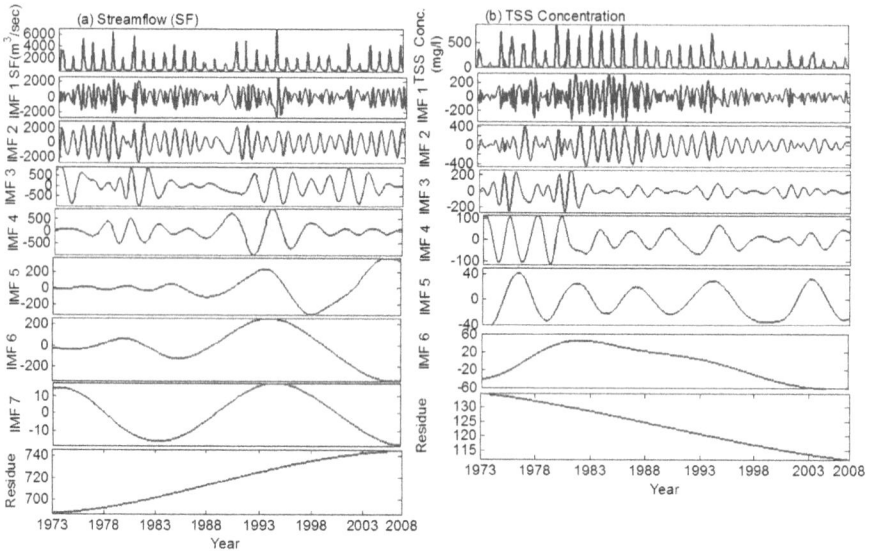

FIGURE 5.2 Decomposition of monthly time series of (a) streamflow and (b) TSS concentration at Basantpur station.

series are presented. Then the results of correlation and TDIC analysis of IMFs of streamflow and sediment concentration data from both the stations are presented.

The CEEMDAN method is applied upon the monthly SF and TSS concentration time series by using a noise parameter of 0.2 and total of 500 realizations to get the ensemble of modes (Torres et al. 2011; Antico et al. 2014). The orthogonal modes obtained for streamflow and TSS concentration time series are presented in Figure 5.2 and Figure 5.3 respectively, for Basantpur and Tikrapara stations.

The decomposition resulted in seven IMFs and a residue for streamflows, while six IMFs and residue for TSS concentration series of Basantpur station. Also, the decomposition resulted in seven IMFs and residue for the dataset of Tikrapara station. The number of modes resulted from the decomposition of all the times series are less than the maximum number expected $\log_2(L)$ (Flandrin and Gonclaves 2004; Flandrin et al. 2004), where L is the length of the dataset. The mean period of IMFs are computed by the zero-crossing method (by counting the number of extrema and hence zero crossing) (Barnhart and Eichinger 2011). The mean period of the modes of streamflow and sediment concentration datasets of Basantpur and Tikrapara stations are presented in Table 5.1. It is also noticed that the annual mode (IMF2) is showing the highest correlation with the respective series for all cases. From the nature of residue, it is noted that both the streamflow and TSS concentration show opposing trend for both the stations during the study period (1973–2007).

Next, the statistical significance test of IMF components (Wu and Huang 2004) of streamflow and TSS concentration data from both stations is performed and the results are presented in Figure 5.4.

From Figure 5.4, it is clear that the annual cycle (IMF2) is significant in all the four time series. In the streamflow and TSS concentration time series, short-term

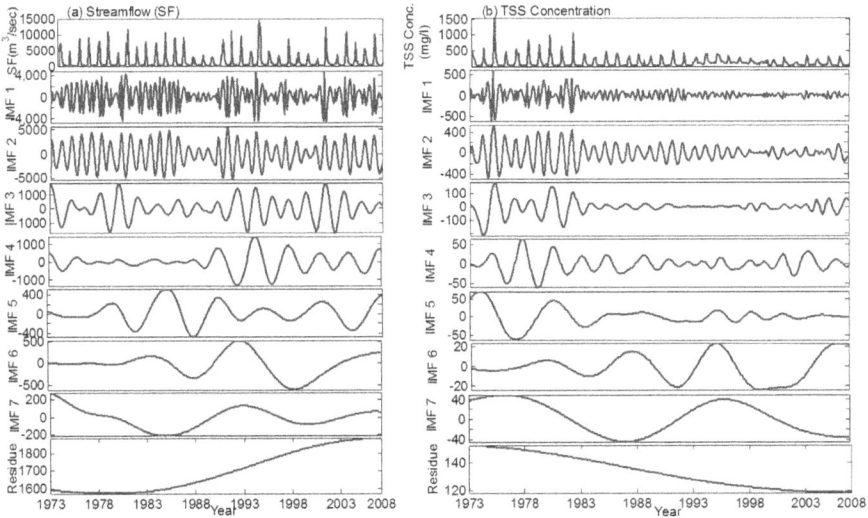

FIGURE 5.3 Decomposition of monthly time series of (a) streamflow and (b) TSS concentration at Tikrapara station.

TABLE 5.1
Mean Periods and Percentage Variability Explained by Different Modes Obtained for Streamflow and TSS Concentration Data from Basantpur and Tikrapara Stations

Mode	Streamflow			TSS Concentration		
	R	T	VE	R	T	VE
(a) Basantpur						
IMF1	0.250	5.771	32.713	0.265	4.810	24.843
IMF2	0.662	11.882	54.713	0.676	10.359	57.605
IMF3	0.193	22.444	6.540	0.289	17.565	10.072
IMF4	0.154	36.727	4.080	0.131	33.667	4.149
IMF5	0.086	80.800	0.978	0.120	80.800	0.987
IMF6	0.005	202.000	0.956	0.209	404.000	2.255
IMF7/Residue	0.025	202.000	0.005	0.161	404.00	0.090
Residue	−0.011	404.000	0.014			
(b) Tikrapara						
IMF1	0.276	5.527	32.374	0.232	4.756	31.224
IMF2	0.675	12.029	58.451	0.683	10.487	57.982
IMF3	0.228	20.450	5.545	0.142	16.360	6.088
IMF4	0.165	31.462	2.251	0.076	25.563	0.805

(continued)

TABLE 5.1 (Continued)
Mean Periods and Percentage Variability Explained by Different Modes Obtained for Streamflow and TSS Concentration Data from Basantpur and Tikrapara Stations

Mode	Streamflow			TSS Concentration		
	R	T	VE	R	T	VE
IMF5	0.074	58.429	0.447	0.052	45.444	1.197
IMF6	0.110	102.250	0.682	0.063	102.250	0.386
IMF7	0.055	204.500	0.122	0.166	204.500	2.021
Residue	−0.018	409.00	0.127	0.097	409.000	0.297

Note: R – correlation coefficient; T – Mean period (month); VE – % Variability explained.

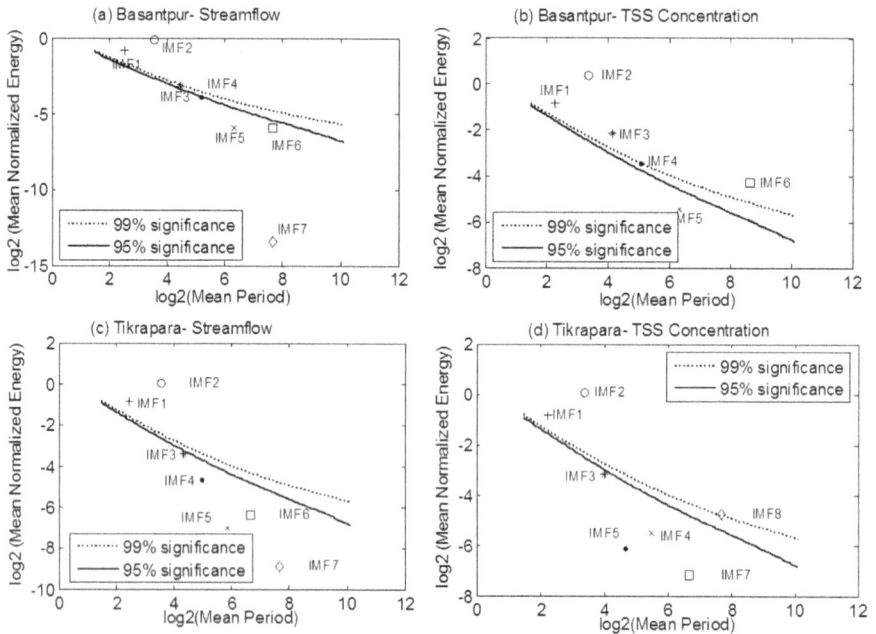

FIGURE 5.4 Statistical significance test of IMF components of streamflow and TSS concentration from Basantpur and Tikrapara stations.

interannual cycles upto 3 years (IMF3 and IMF4) are significant. But in the stream-flow variability of Tikrapara station, none of the IMFs other than annual cycle (IMF2) is found to be significant and for TSS series also the energy level of IMF4 and IMF6 is relatively less for the Tikrapara station when compared with that of Basantpur station. This may be attributed to the fact that the station is located at far down-stream of the major control structure – the Hirakud dam and human interventions are more influential in the variability of streamflow and TSS time series of Tikrapara

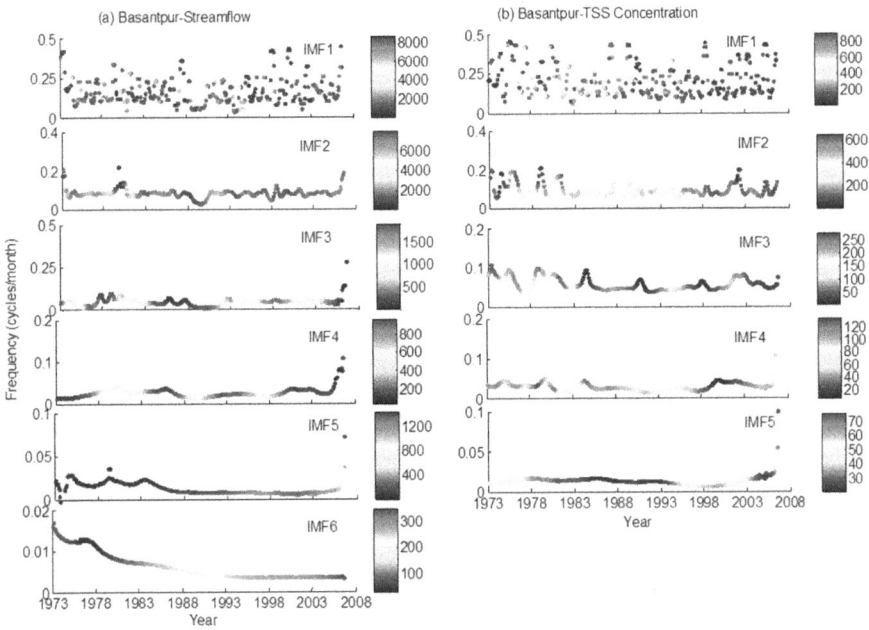

FIGURE 5.5 Instantaneous frequency trajectories of (a) streamflow (b) TSS concentration data from Basantpur station.

station. Further, the obtained IMFs are subjected to the NHT-DQ method to get the instantaneous amplitudes and frequencies. The Hilbert spectral representations of different IMFs are presented in Figure 5.5 and Figure 5.6, respectively, for the data from Basantpur station and Tikrapara station.

From the spectral representation of different data (Figure 5.5 and Figure 5.6) some common features can be identified. All the time series shows more intermittency in IMF1 (~5- to 6-month periodicity). The highest amplitudes of high frequency modes are found to be localized in time. The dominant frequency for different IMFs (at which there is a concentration of higher amplitudes) is found to be occurring at different time instant, i.e., dominant frequency is time varying in characteristics. This time varying property of sediment concentration time series was also reported by Cao et al. (2007) and Kuai and Tsai (2012) in earlier studies.

The mean instantaneous frequencies of first five IMFs of streamflow time series of Basantpur station are 0.166, 0.085, 0.043, 0.027 and 0.012; and the corresponding mean period are 6.03, 11.73, 23.09, 37.35, 81.08, which are quite similar to the periodicity based on zero crossing method, particularly for the lower modes. For the low frequency modes, the periodicity may differ from that obtained by zero crossing method and the dyadic behavior of periodicity may not get preserved because of a lower number of cycles present in such modes. From Figure 5.5 and 5.6, the highest variability in instantaneous frequency is for IMF1. This component shows high turbulence of instantaneous frequencies and it is responsible for non-linearity of the dataset, as noticed by Kuai and Tsai (2012) in a similar study. Also for all the four time series, the mean amplitude is maximum for IMF2, which denotes near annual periodicity.

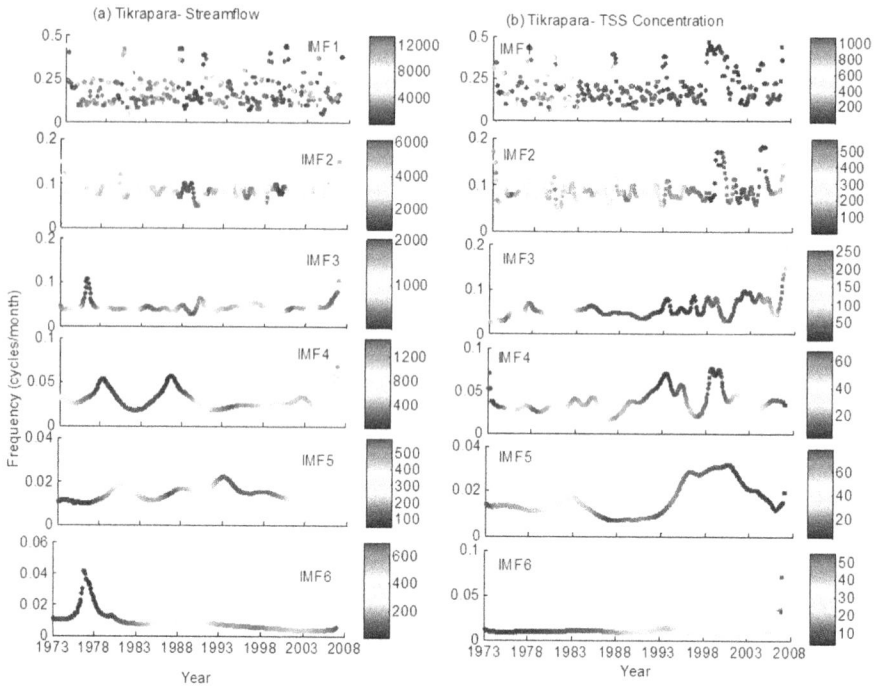

FIGURE 5.6   Instantaneous frequency trajectories of (a) streamflow (b) TSS concentration data from Tikrapara station.

Further, the marginal spectrum of the four time series (streamflow and TSS concentration of both stations) are prepared and presented in Figure 5.7. From the different panels of Figure 5.7, it is noticed that both Fourier and Hilbert spectral analysis identified a prominent spectral period near annual scale (~0.085 cycles/months). Furthermore, it is noticed that for mean instantaneous frequencies between 0.012 months to 0.085 cycles/months (corresponding to period of 7 years to 1 year), the evolution of spectra are quite similar for streamflow and TSS concentration data of Basantpur station. This could be referred as climatic regime where inter annual climate oscillations may influence the variability of streamflow (Thompson and Katul 2012). Hence this could be possibly linked with the large-scale climatic oscillations operating at that scales which control the variability of streamflow and TSS concentrations of Mahanadi basin. Gadgil et al. (2004) highlighted that two such large scale circulation patterns from Pacific Ocean and tropical Indian Ocean (ENSO-ElNiño Southern Oscillation, and EQUINOO –Equatorial Indian Ocean Oscillation) operating at similar scales significantly influence the precipitation processes in the Indian subcontinent; and Maity and Nagesh Kumar (2008) noticed that such circulation patterns can influence the streamflow variability of Mahanadi basin. However, there is difference in the evolution of two spectra of Tikrapara station which is located at far downstream of Hirakud reservoir. It is logical to ascertain that the streamflow and sediment concentrations of Basantpur are influenced more by

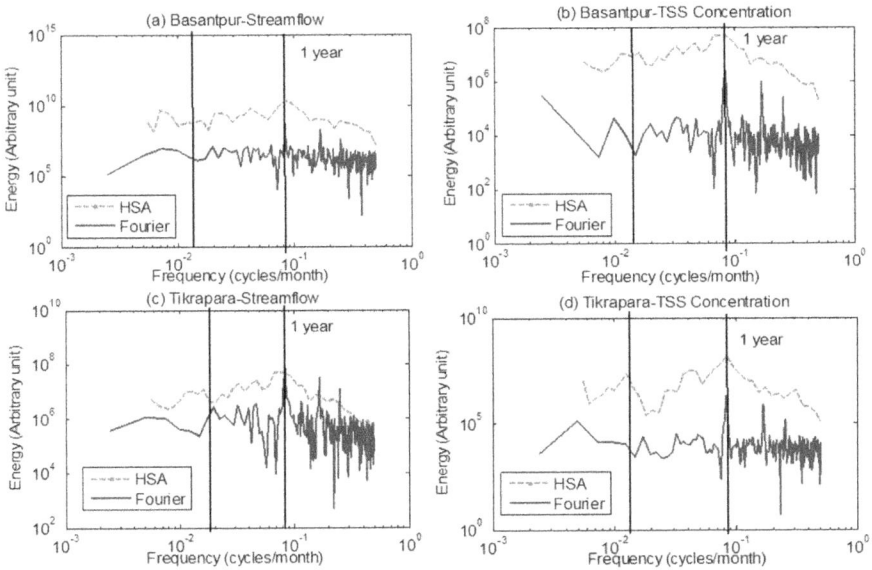

FIGURE 5.7 Marginal Hilbert spectrum and Fourier spectrum for streamflow and TSS concentration. (a) and (b) for data from Basantpur station; (c) and (d) for data from Tikrapara station. The right vertical bar refers to annual time scale while the left vertical bar refer the interannual time scale.

such external forcing, while that at Tikrapara could be influenced by internal factors like basin characteristics and human interventions.

5.2.3 INVESTIGATING THE STREAMFLOW-SEDIMENT LINK USING TDIC ANALYSIS

Investigating the strength and nature of association between streamflow and sediment concentration data is a popular hydrologic problem, for which different methods such as simple statistical correlations, scanning F-test, t-test, coherency analysis, continuous wavelet transforms were used in the past (Zhang et al. 2008, 2009, 2012; Chen et al. 2010b). By investigating such linkages, the reason for hydrological variability of the river basin can be interpreted (like climatic oscillations or human interventions). The procedure of multiscale investigation of hydroclimatic teleconnections described in Chapter 3 can also be effectively used for investigating the link between the hydrological variables such as streamflow and sediment concentration. The complete procedure involving TDIC analysis for investigating the streamflow-sediment link is depicted as flowchart in Figure 5.8.

A visual comparison of modes and quantitative estimation of correlation between different modes enable us to find the association between the streamflow and TSS concentration time series at different scales over different time spells. Therefore, first a plot of IMFs and residue of both streamflow and TSS concentration time series from Basantpur station are made and presented in Figure 5.9. A careful perusal of

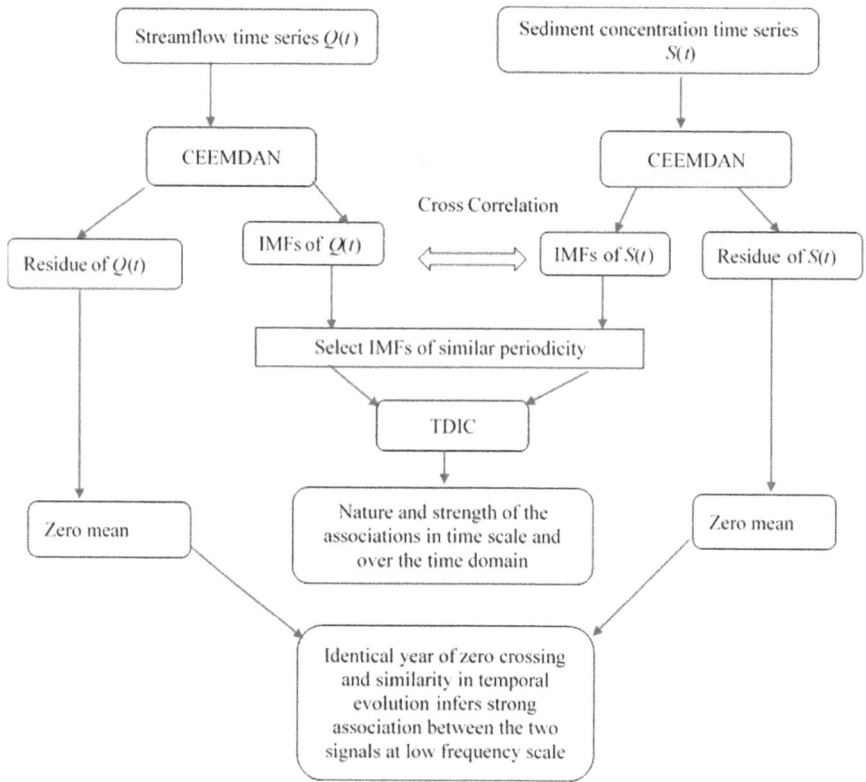

FIGURE 5.8 Flowchart of CEEMDAN-TDIC approach for investigating streamflow-sediment links.

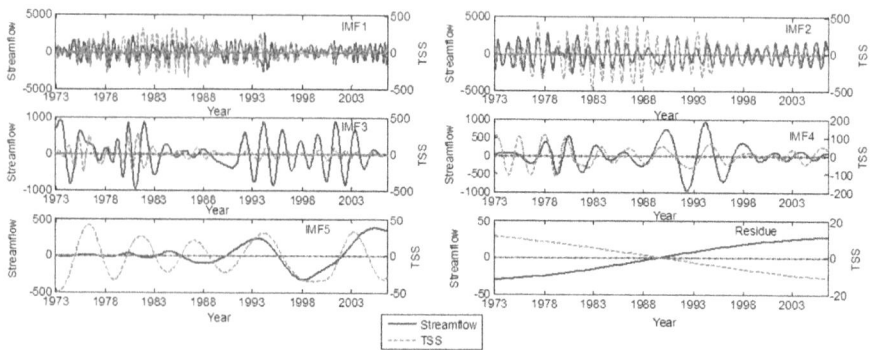

FIGURE 5.9 Comparison of first five IMFs and the residue of streamflow with that of TSS concentration time series of Basantpur station.

the plots of IMFs shows that during most of the time periods, the evolution of IMFs of TSS matches well with that of streamflow, particularly in high frequency modes. It is noticed that a strong direct in-phase relation exists between the annual modes of streamflow and TSS concentration. However, in some of the time spells, weak relationship between the IMFs is also noticed (for example, until 1990s in IMF 5, before 1980s in IMF4). The trend component (residue) is a slowly varying climate mode, which delineates the progressive change from below average to above average streamflow (or TSS). From Figure 5.9, it is also noticed that the temporal evolution of trend (i.e., how the trend changes over and below the zero mean) components of SF and TSS concentration are similar for most of the periods, and the zero crossing is happening nearly at the same year of 1990. The plots of IMFs of streamflow and TSS concentration time series from Tikrapara station is also made and presented in Figure 5.10. Here also in the high frequency IMFs, there is an in-phase relation between the two but in the low frequency modes (IMF3-IMF5) there is a clear dominancy of anti-phase (negative) relation between streamflow and TSS in the time domain.

To quantitatively ascertain the association between streamflow and TSS concentration, first the overall correlation between the TSS and streamflow data is computed, and it is found to be 0.76 at Basantpur station; and 0.69 at Tikrapara station. But these values only give a linear association between the two time series, which alone is not sufficient to find the true association between two non-linear and non-stationary time series. A detailed cross correlation analysis is made between the orthogonal components of streamflow and that of TSS concentration series of both Basantpur and Tikrapara station and the results are presented in Table 5.2.

In Table 5.2, it can be seen that there exists strong positive association between corresponding IMFs of the series (between IMF1 of streamflow and that of TSS, IMF2 of streamflow and that of TSS, etc.), but there exists a strong negative correlation between the two (−0.99) for the residue. Similar observation can be made for other IMFs also. Similarly, strong positive correlation between the respective modes is noticed for all modes except IMF4 and the residue in the decomposition of data from Tikrapara station. Therefore, the possibility of a reversal in the nature of association between the two series in different scales and along the time domain cannot be ignored. The negative correlation between sediment load and streamflow on longer time scales confirm the general trend of sediment load and streamflow (Zhang et al. 2009). Also, it is to be noted that the above cross correlation analysis is done by considering the entire time span. Based on the comparison of different IMFs (Figure 5.9 and Figure 5.10), it can be inferred that if shorter time spells are considered, an opposing nature of correlation may exists between streamflow and TSS irrespective of the nature of overall correlation between different IMFs. In addition, overall small correlation between streamflow and sediment at certain time scales (for example, IMF3 in Basantpur station and IMF4 in Tikrapara station) might be because of the existence of such opposing correlations for shorter time spells. Such opposing nature of correlation may nullify the association and finally lead to the quite smaller overall correlation in that time scale. Hence it is very important to identify the period of existence of such unusual negative (positive) correlation between the two and the periodic time scale at which such correlation exist. To get information on such local association between the two series and the strength of the

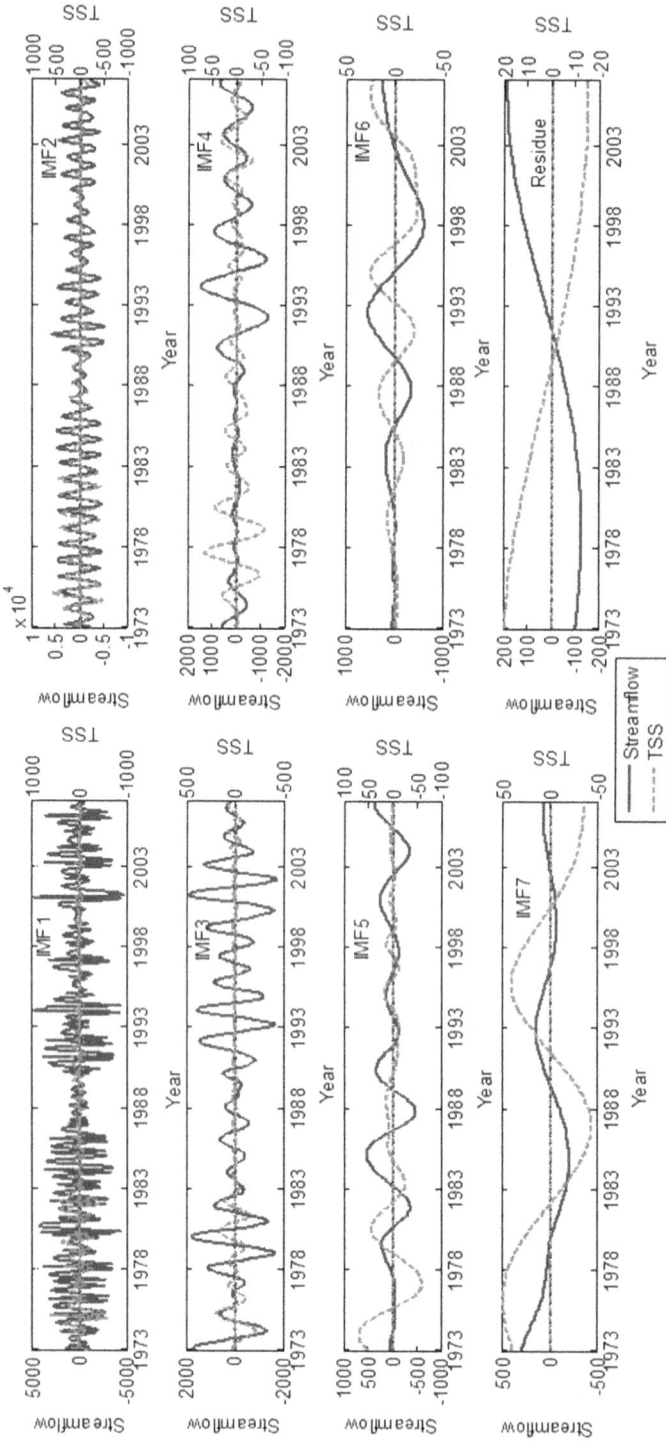

FIGURE 5.10 Comparison of oscillatory modes of streamflow with that of TSS concentration time series of Tikrapara station.

TABLE 5.2
Correlation Coefficients Between Orthogonal Components of Streamflow and TSS Concentration at Basantpur Station and Tikrapara Stations

TSS Concentration	Streamflow							
	IMF1	IMF2	IMF3	IMF4	IMF5	IMF6	IMF7	Residue
(a) Basantpur								
IMF1	**0.524**	–0.258	0.033	0.006	–0.033	–0.007	–0.042	–0.047
IMF2	–0.135	**0.648**	0.016	–0.038	0.016	–0.025	–0.005	0.028
IMF3	–0.196	0.390	0.076	0.002	0.005	0.004	0.013	0.011
IMF4	–0.064	0.016	0.180	**0.550**	0.026	–0.082	–0.010	0.004
IMF5	–0.025	0.012	0.034	–0.128	0.327	0.076	–0.038	–0.093
IMF6	–0.009	–0.009	–0.069	–0.018	–0.185	0.431	–0.174	**–0.636**
Residue	0.015	–0.021	0.087	–0.008	–0.187	0.242	0.072	**–0.999**
(b) Tikrapara								
IMF1	**0.558**	–0.354	–0.019	–0.017	–0.011	0.028	0.024	0.028
IMF2	–0.270	**0.774**	0.026	0.035	–0.012	–0.035	0.003	0.018
IMF3	–0.052	0.023	0.300	0.229	0.000	0.015	0.014	0.022
IMF4	0.008	–0.003	0.069	0.019	–0.001	0.062	–0.013	–0.011
IMF5	0.014	0.005	0.031	–0.044	–0.038	–0.125	–0.057	0.000
IMF6	0.016	–0.021	0.005	0.044	–0.128	0.200	–0.083	0.146
IMF7	–0.051	0.006	–0.128	0.042	0.213	0.109	**0.556**	0.060
Residue	–0.007	0.028	0.056	0.042	–0.058	–0.045	–0.264	**–0.951**

Note: Bold numbers indicate significant correlation at 5 % significance level.

association, a running correlation analysis framed upon sliding windows is to be used. The HHT-based TDIC analysis is one such multiscale running correlation analysis, which is helpful to get such information on local correlations. So the TDIC analysis is performed between the IMFs of streamflow and TSS concentration with comparable periodicities. The TDIC plots are prepared for different IMFs and presented in Figure 5.11 and Figure 5.12, respectively, for Basantpur and Tikrapara stations.

The TDIC analysis between streamflow and TSS concentration in Basantpur station showed a dominancy of strong long-range positive correlation in IMF1, IMF2 and IMF4. However, in IMF3 and IMF5 a negative association between the two is noticed during ~1980–1990 period. More importantly, in the above modes, the nature of association changes over the time domain, i.e., there are frequent reversals of correlation from positive to negative (and vice versa) particularly at intra decadal time spells (say, less than 100 months). The positive association (in-phase relationship) between sediment load and streamflow seems to indicate the dynamics of hydrological processes on the transport of suspended sediment in the river channel (Milliman and Syvitski 1992). Variations in the discharge characteristics of river flow determine the transport and suspended sediment fluxes and can help in understanding the factors and processes influencing the sedimentation in rivers. In general, the greater the flow,

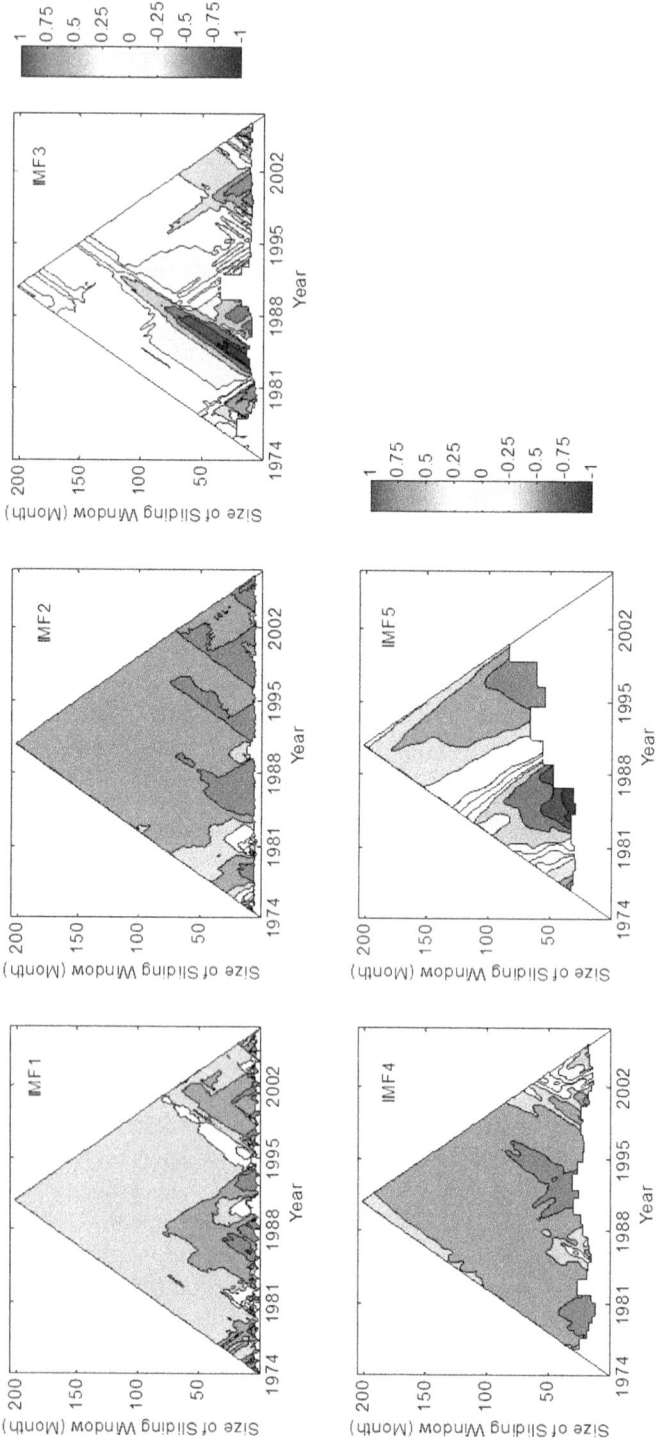

FIGURE 5.11 TDIC plots between IMFs of streamflow and TSS concentration data from Basantpur station. The white space of the TDIC plot means that the correlation coefficient is not significant at 5% significance level.

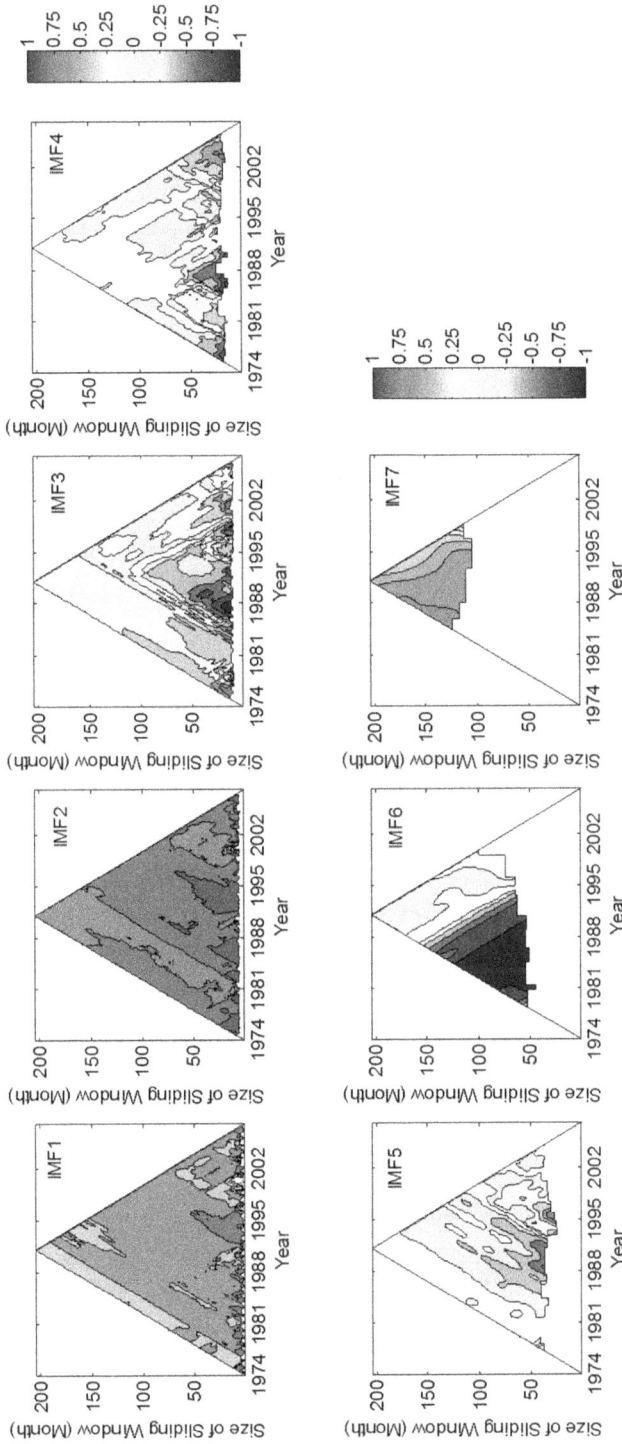

FIGURE 5.12 TDIC plots between IMFs of streamflow and TSS concentration data from Tikrapara station. The white space of the TDIC plot means that the correlation coefficient is not significant 5% significance level.

the more sediment will be transported. Water discharge can be strong enough to suspend particles in the water column as they move downstream, or push them along the bottom of a water course (Sadeghi and Mostafazadeh 2016). The negative relation between sediment load and streamflow implies that sediment load or streamflow was heavily influenced by other external factors such as human activities, which include flow diversion, construction of check dams, etc., along with the construction of major control structures (Zhang et al. 2008). TDIC analysis performed for the streamflow and TSS concentration data from Tikrapara station (Figure 5.12) shows similar trend in IMF1 and IMF2. Long-range weak positive correlation between the two series is noticed in the interannual mode IMF3, with frequent reversals in the nature of correlation during the period 1980–1990. Unlike for the case of Basantpur station, in this station the modes are negatively correlated in long range in the low frequency modes IMF4 and IMF5. The strong negative association between streamflow and sediment concentration is more perceptible in the low frequency IMFs of the time series from Tikrapara station than that from Basantpur station. This could be attributed to the fact that the location of Basantpur is at upstream of major control structure Hirakud dam, while the location of Tikrapara is at far downstream of Hirakud dam. The influence of human interventions in the region might also be a reason for such negative association between streamflow and TSS concentration. From the TDIC analysis, it is found that the nature and strength of association between the streamflow and sediment is fairly consistent at annual and intra-annual scales. But, in few inter annual modes, there can be frequent alterations in the nature of correlation between the two. Eventhough an obvious reasoning is not possible to be adduced in this regard by the present study, it can be speculated that such transition in correlation may be attributed to some unidentified physical processes, the spatial nonhomogenity of the basin or the influence of different climate forcing on the local hydrological processes.

5.2.4 MULTIFRACTAL DESCRIPTION OF DAILY STREAMFLOW AND TSS DATASETS USING AOHSA

The daily streamflow time series are often multiscaling in character, also characterized by high turbulence are well proven to be multifractal in nature (Pandey et al. 1998; Kantelhardt et al. 2006; Li et al. 2015). It is well proven that raising the Hilbert energy spectrum to statistical moments of arbitrary order, one can investigate the multifractality of the time series (Huang et al. 2009a). However, the practical applications of the so-called Arbitrary Order HSA method for multifractal analysis being scarce in literature, this study investigated its potential for AOHSA method for description of multifractality of streamflow and TSS concentration data of Basantpur and that at Tikrapara stations. Daily records of streamflow and TSS concentration at Basantpur and that at Tikrapara for the period 1973–2007 collected from Central Water Commission (www.india-wris.nrsc.gov.in/) are used for the present study and the time series are presented in Figure 5.13.

The daily streamflow and TSS concentration time series are first decomposed adaptively by the CEMDAN method. The decomposition resulted in 20 and 15 modes for the streamflow and TSS concentration data from Basantpur station, while it resulted in 20 and 16 modes for streamflow and TSS concentration data from

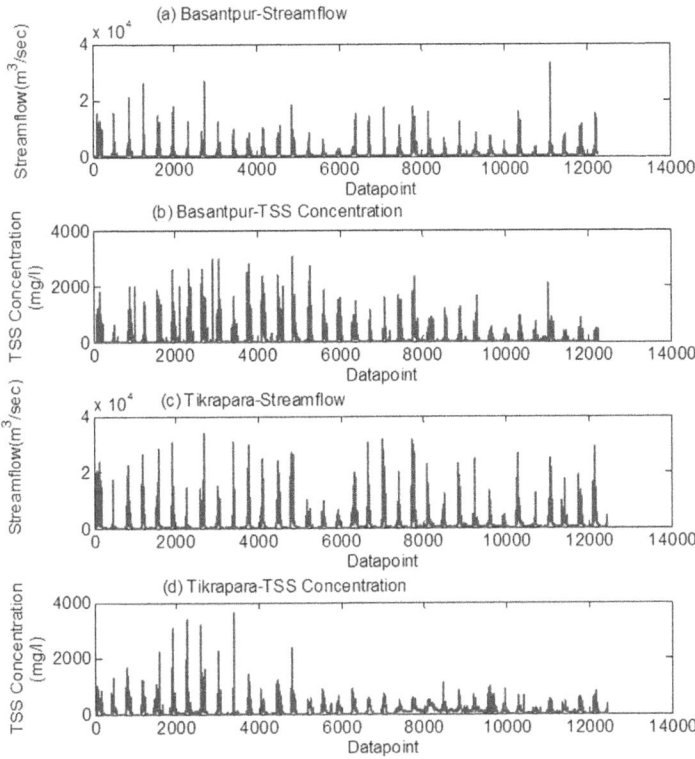

FIGURE 5.13 Daily time series of streamflow and TSS concentration data from Basantpur and Tikrapara stations.

Tikrapara station. For brevity, the modes of decomposition results are not presented here. However, the basic properties of the IMFs are verified and analyzed. Both the length of the dataset and the complexity of the dataset plays a role in deciding the number of modes from the decomposition (Flandrin et al. 2004; Huang et al. 2009a). Then the modes are subjected to Hilbert transformation and mean frequencies are computed. The estimates of the mean frequencies (corresponding to different IMFs of different time series) are plotted against the number of modes in log-linear representation and presented in Figure 5.14.

It is found that mean frequency follows an exponential law with the mode number as $\bar{\omega} = \gamma^{-n}$, where n is the index of modes and γ is the scaling exponent, which implies that the mean frequency of a given mode is γ times larger than the mean frequency of the next one. The estimates of γ are found to be centered around 1.5 for different series (1.4486, 1.5373 for streamflow and TSS concentration series from Basantpur station; 1.4229, 1.4597 for the streamflow and TSS concentration series from Tikrapara station. It clearly shows that the scaling exponents are quite different from 2, the ideal dyadic filter bank property of modes. This also influence the number of modes evolved where the number of modes obtained cannot be close to $\log_2(N)$ in such cases where N is the data length (a case for dyadic filter banks). But the power

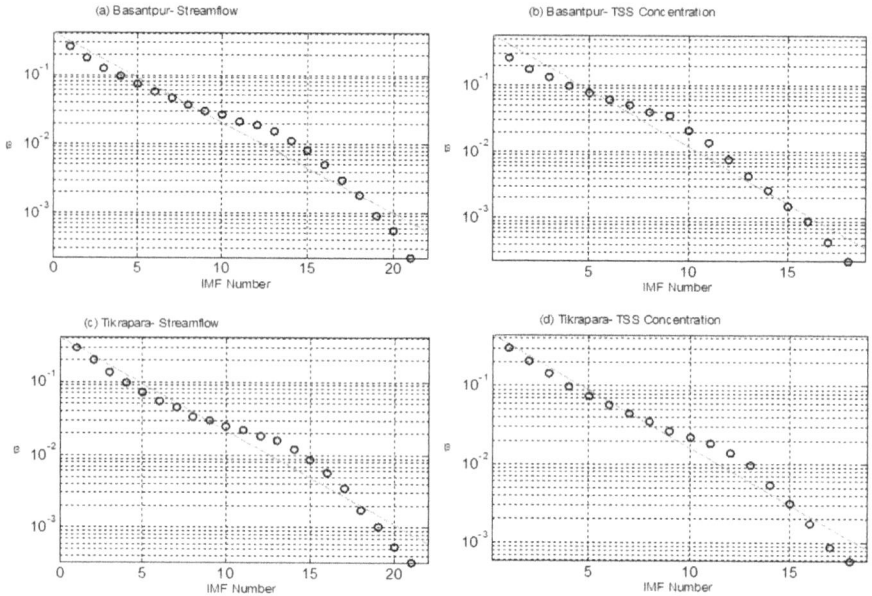

FIGURE 5.14 Representation of mean frequency vs mode index of IMFs of streamflow and TSS concentration time series from two stations.

law variation suggests that EMD still acts as a filter bank. Huang et al. (2009a) also noticed such property while analyzing the daily streamflow of two rivers from France in earlier studies. To characterize the intermittency properties, first the Hilbert energy spectrum (for order 2) and Fourier spectrum of different time series are prepared. The spectra for different stations are presented in Figure 5.15.

The visual examination of the spectra clearly shows that both the Fourier and HSA method detects a prominent peak at frequency of ~0.0028 (i.e., a period of 357 days). Thus, it could be ascertained that the annual periodicity is clearly detected by the Fourier and HSA method in different time series. The Hilbert energy spectra of the different time series are presented in Figure 5.16, which clearly show that the scale invariance holds at intra-annual scale ranges.

The vertical bars marked in Figure 5.16 indicate the scale invariance range, it could be stated that the scale invariance holds in the range $0.4 < \omega < 0.01$ day^{-1} approximately (corresponding to a time scale of 2.5 days to 100 days) for different time series, which is typical time scales for the turbulence property (Thompson and Katul 2012). Hence some researchers believe that the hydrological process is an analogy with the stochastic cascade models in a fully developed turbulent flow that generally yield fractals (Shang and Kamae 2005; Kuai and Tsai 2012). Hence one can argue that varying time scales in hydrological time series may display the self-similarity and it is a signature of fractals. Kuai and Tsai (2012) proved that the turbulence property in sediment and streamflow time series is attributed to the fractal character based on the computation of slope of Fourier power spectrum.

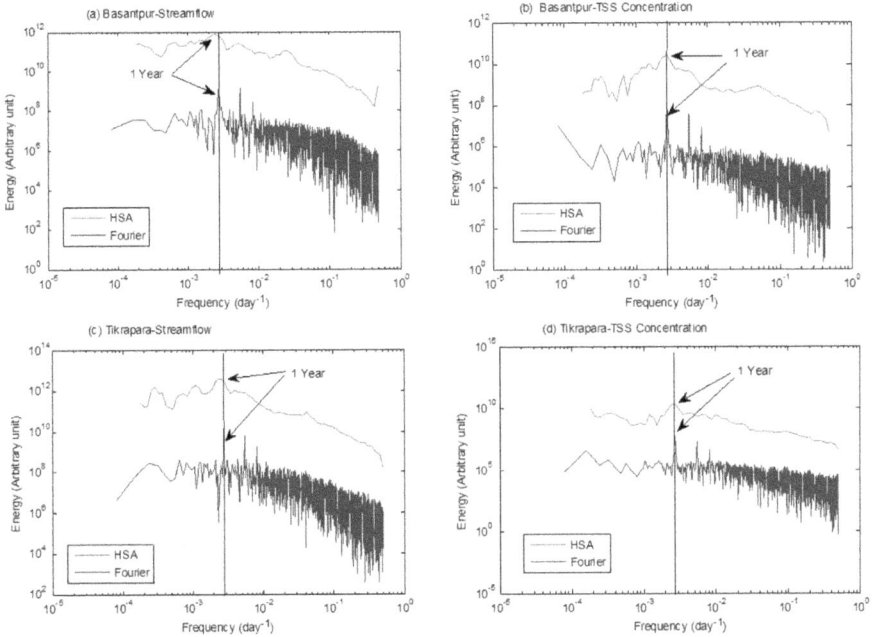

FIGURE 5.15 Comparison of Fourier spectra and marginal Hilbert spectra of streamflow and TSS concentration time series from Basantpur and Tikrapara stations.

Arbitrary Order Hilbert Spectral Analysis (AOHSA) proposed by Huang (2009), is an extension of traditional HSA for q order moments. Huang et al. (2009a) demonstrated its usefulness for detection of multifractality of streamflow time series from Seine and Wimereux rivers in France. HSA of the different time series considered in this study proven that the scale invariance persists in those time series at intra annual scale ranges (varying between 3 days to 3 months approximately). For different moment orders q = 0 – 3, the arbitrary order Hilbert spectra are prepared and presented in Figure 5.17.

The results show that the spectra for moment order 3 is slightly irregular with high fluctuation in spectral amplitude values. The slopes of the spectra for different moment order q give the scale exponent $\xi(q)$, which enables us to make a plot of $\xi(q)$ vs q. This plot is given in Figure 5.18. The concavity of this plot infers multifractality of the respective time series. The appropriate detection of scale range is a crucial task in this method. The large fluctuations of spectral amplitudes for the spectra of higher orders (say $q > 3$) may introduce error in estimation of scaling exponent. Therefore this study consider the moment orders up to three only, which supports the recent findings of Lombardo et al. (2014) that up to moment order three is sufficient to describe the multifractality of hydrologic time series.

The quantity $\xi(1) - 1$ is an indication of short/long memory dependency structure of the time series and the quantity is similar to the classical Hurst exponent (H) (Huang et al. 2009a). The values of Hurst exponent (H) computed from the scaling exponent

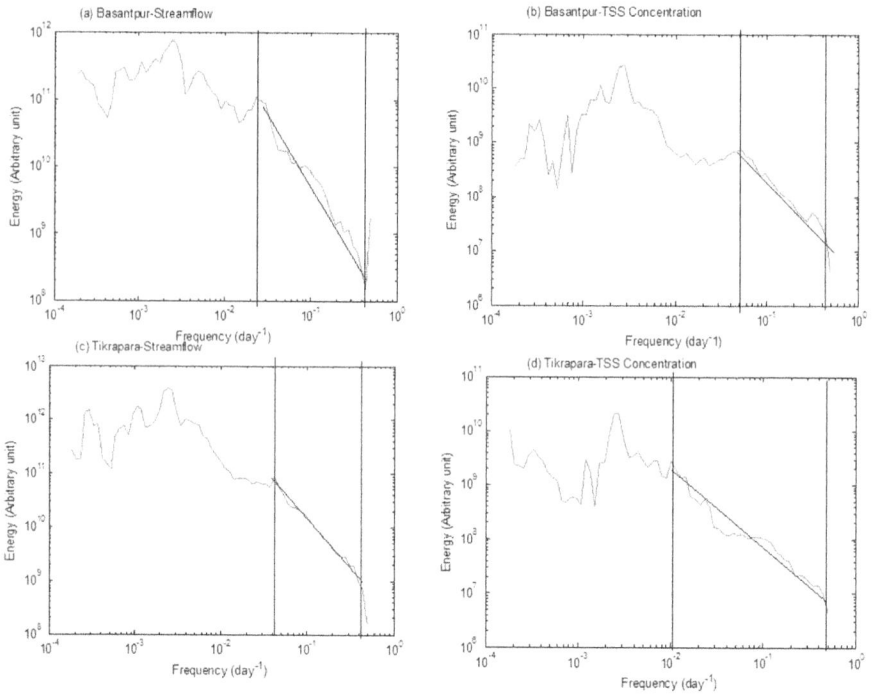

FIGURE 5.16   Hilbert energy spectra of streamflow and TSS concentration time series from Basantpur and Tikrapara stations. The vertical bars indicate scaling ranges.

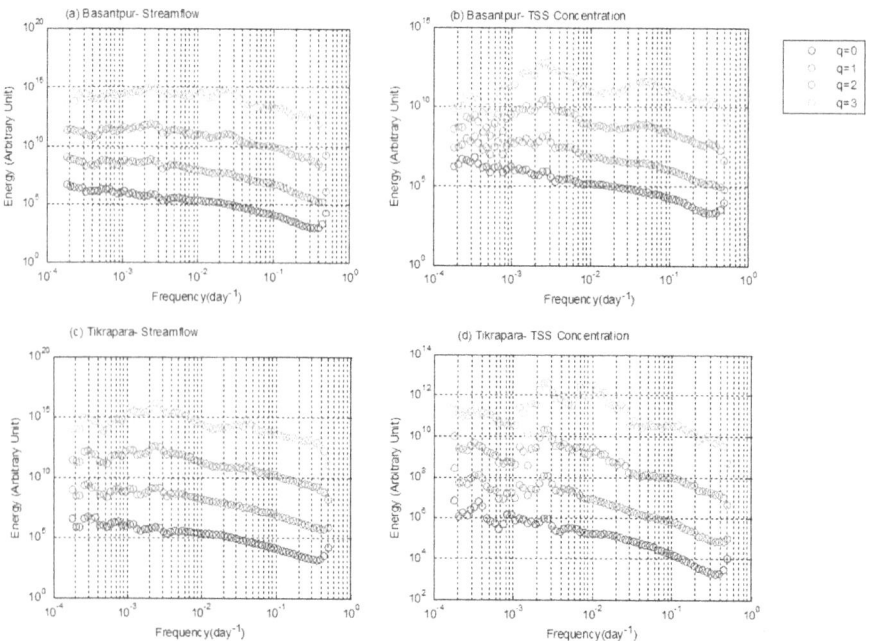

FIGURE 5.17   Marginal Hilbert spectra of streamflow and TSS concentration series from Basantpur and Tikrapara stations for different moment orders.

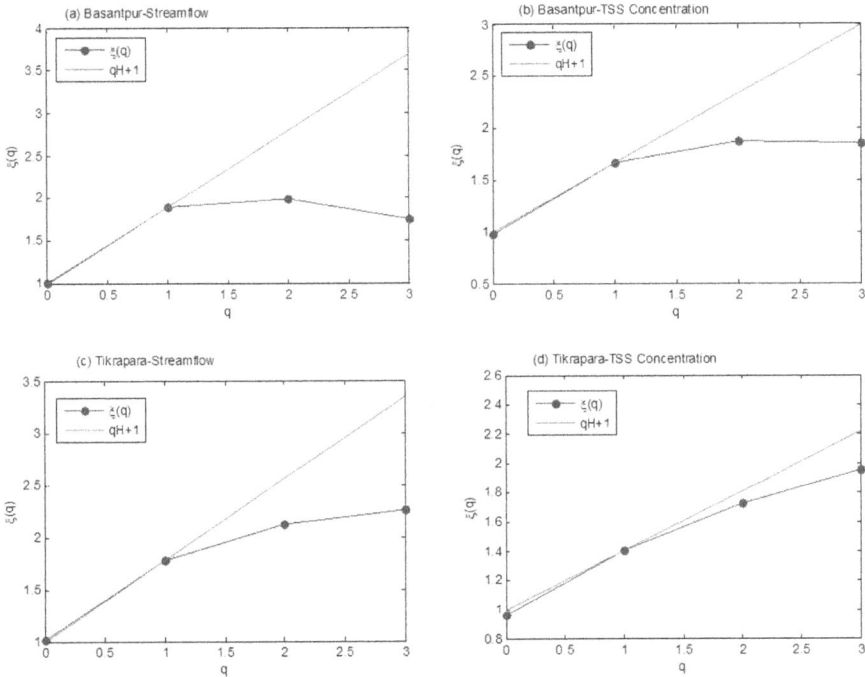

FIGURE 5.18   Scaling exponents of streamflow and TSS concentration series from Basantpur and Tikrapara stations for different moment orders. Here $\xi(q)$ is the scaling exponent in Hilbert space.

in Hilbert space ($\xi(q)$) are 0.89, 0.665, 0.782 and 0.406 for different cases. All four time series are indicating the long term dependency structure of the series, except for TSS Concentration series of Tikrapara station. From Figure 5.18, it is also inferred that the curve becomes flat or shows a reduction beyond the order $q=1$, for the data from Basantpur station. But for Tikrapara station it is more concave in nature. This indicate that there is less degree of multifractality (and hence less intermittency) for the records from Tikrapara station.

5.3   MEMD–TDIC COUPLED FRAMEWORK FOR INVESTIGATING MULTISCALE TELECONNECTIONS OF RESERVOIR INFLOW

5.3.1   DATA DESCRIPTION

Hirakud reservoir is a multipurpose reservoir located on Mahanadi river in the state of Orissa in India, having a catchment area of 83,400 sq. km. The operation of Hirakud reservoir plays an important role on the irrigation, hydropower production and flood control over downstream area. The uncertainty of inflows into Hirakud reservoir has high influence on the operation of the reservoir system. The streamflows during the monsoon period (June to September) constitutes most of the annual inflow into Hirakud reservoir, which shows a faster variation, and inflows during the nonmonsoon period show a slow variation. To demonstrate the applicability of HHT for multiscale spectral analysis of reservoir inflow, this study uses the monthly inflow time series of Hirakud reservoir in Mahanadi river, India for 36 years (1958–1993). To investigate the influence of large-scale circulations on the variability of Hirakud reservoir inflows for the above stated period, ENSO and EQUINOO indices are considered in this study. As the SST anomaly corresponding to the Niño 3.4 region (120°W–170°W, 5°S–5°N) is well linked with the rainfall over Mahanadi (Maity and Nagesh Kumar 2009) the values of SST for Niño 3.4 region obtained from NOAA National Weather Service Climate Prediction Center (www.cpc.ncep.noaa.gov/data/indices/) is used as representative data of ENSO. Also the negative of the anomaly of the zonal component of surface wind in the equatorial Indian ocean region (60–90°E, 2.5°S–2.5°N) is considered as EQUINOO index (Maity and Nagesh Kumar 2006a). The zonal wind component for the same period (1958–1993) was obtained from the National Centre for Environmental Prediction (NCEP) www.esrl.noaa.gov/psd/data/gridded/data. ncep.reanalysis.html).

5.3.2 HHT ANALYSIS OF RESERVOIR INFLOW BASED ON MEMD

In this study, the monthly inflow time series of Hirakud reservoir along with the two potential climate variables ENSO and EQUINOO constitute the multivariate dataset. For the implementation of MEMD algorithm, the minimum and maximum threshold parameters are set as 0.075 and 0.75, respectively, while the fraction controlling the sifting iterations is fixed as 0.075, following the recommendation from the past studies (Hu and Si 2013). The decomposition resulted in six modes and residue, which is less than the expected maximum of $\log2(N)$, N is the data length (420) (Flandrin et al. 2004; Huang et al. 2009b). The obtained orthogonal modes are presented in Figure 5.19. The mean period of the different modes are computed by 'zero crossing' method (Barnhart and Eichinger 2011). The mean period of each mode and the percentage variability explained by different IMFs is given in Table 5.3. It is observed that for first few modes, the mean period is showing a dyadic property and the dyadic nature gets violated for the higher modes, because of only few cycles are present in the higher order modes. Hence, it can be concluded that eventhough IMFs act as a filter bank, it need not be truly dyadic (Huang et al. 2009a). This can be considered as a positive feedback, as unlike the case of a discrete wavelet-based multiresolution analysis (which performs the decomposition such that mean period in powers of 2, for different modes), EMD identify the true scale of the processes involved. HHT is capable of detecting the actual periodic scale inherent in the dataset. From Figure 5.19, it is noticed that, IMF4 of inflow shows very low variability for the period (1964–1976). It is also noted that during the time domain under consideration (1958–1993),

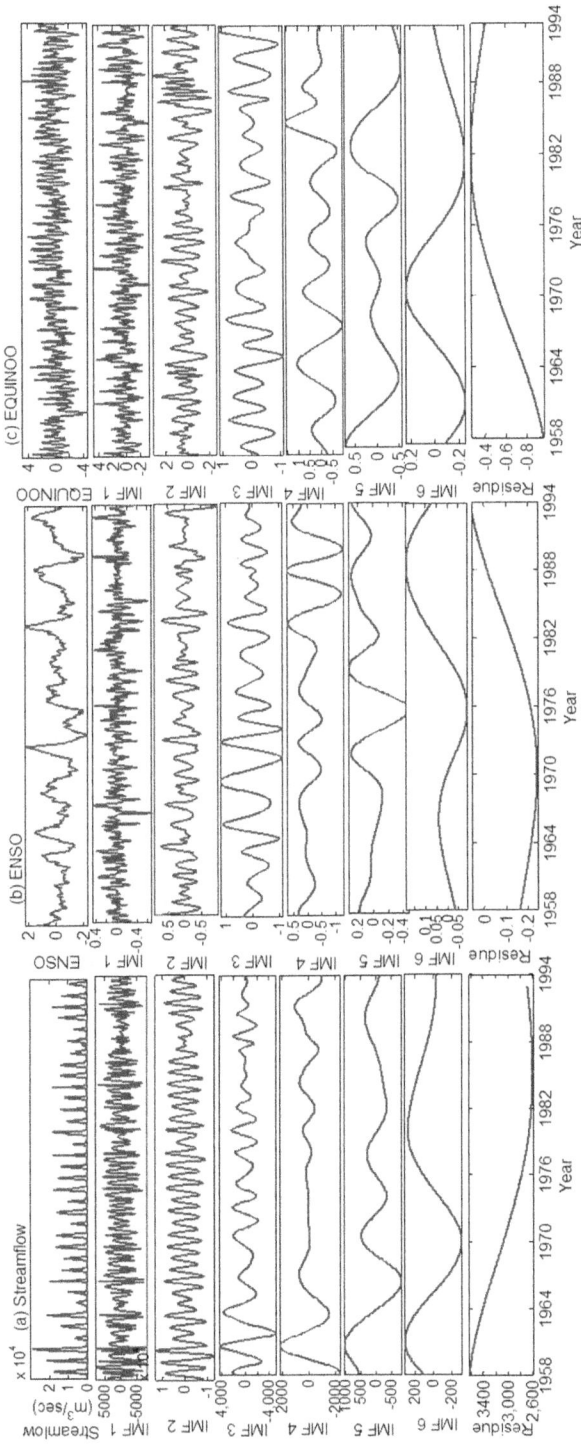

FIGURE 5.19 Orthogonal modes of (a) monthly inflow into Hirakud reservoir; (b) ENSO and (c) EQUINOO time series for 36 years (1958–1993). (From Adarsh 2017).

TABLE 5.3
Mean Period and Percentage Variability Explained by the Modes of Inflows and Climate Oscillations Time Series

Mode	Inflow		ENSO		EQUINIOO	
	Mean Period (in months)	% variability explained	Mean Period (in months)	% variability explained	Mean Period (in months)	% variability explained
IMF1	5.52	27.92	4.62	4.95	5.33	54.64
IMF2	11.57	63.95	14.03	13.90	10.67	26.75
IMF3	25.85	5.85	28.52	41.21	27.57	5.42
IMF4	59.85	1.38	66.68	32.04	62.73	7.45
IMF5	118.22	0.49	125.54	5.21	106.56	3.66
IMF6	279.39	0.09	255.58	1.26	281.01	0.68
Residue	LT	0.33	LT	1.44	LT	1.41

LT refers to long term.

the reservoir inflow shows a decreasing trend while the climate oscillations show an increasing trend.

The statistical significance test of IMF components (at 5% significance level) is performed following the procedures described by Wu and Huang (2004). The results of statistical significance test is presented in Figure 5.20, and it can be noticed that IMF2 and IMF4 (i.e., annual and interannual modes of ~2 and ~4 years) are statistically significant.

Different IMFs obtained by MEMD are subjected to Hilbert transformation to find instantaneous frequencies and instantaneous amplitudes. For brevity, the instantaneous frequency trajectories in a time-frequency domain are not provided, but the integrated MHS is presented in Figure 5.21 and it is compared with Fourier power spectrum. From Figure 5.21 it is observed that both the methods are successful in detecting the spectral peak with near annual cycles (frequency ~0.085 month^{-1}). Figure 5.21 further shows that in the range of frequency of ~0.085 cycles/month to 0.01 cycles/month (corresponds to annual to interannual scale of ~8 years), the evolution of both the spectra are similar. This range of frequency corresponds to the climatic regime and possible association of hydrologic variables with climatic drivers could be suspected (Thompson and Katul 2012). This shows that reservoir inflow could be linked with large scale circulations such as ENSO and EQUINOO of similar periodic scale. Therefore, the teleconnection between the reservoir inflows and two large scale circulations (ENSO and EQUINOO) could be investigated in detail with the help of TDIC analysis in a multiscaling framework.

5.3.3 TDIC ANALYSIS BETWEEN RESERVOIR INFLOW AND LARGE-SCALE CLIMATE OSCILLATIONS

To investigate the association between the large scale circulations (ENSO and EQUINOO indices) and reservoir inflow, first the cross correlation analysis between

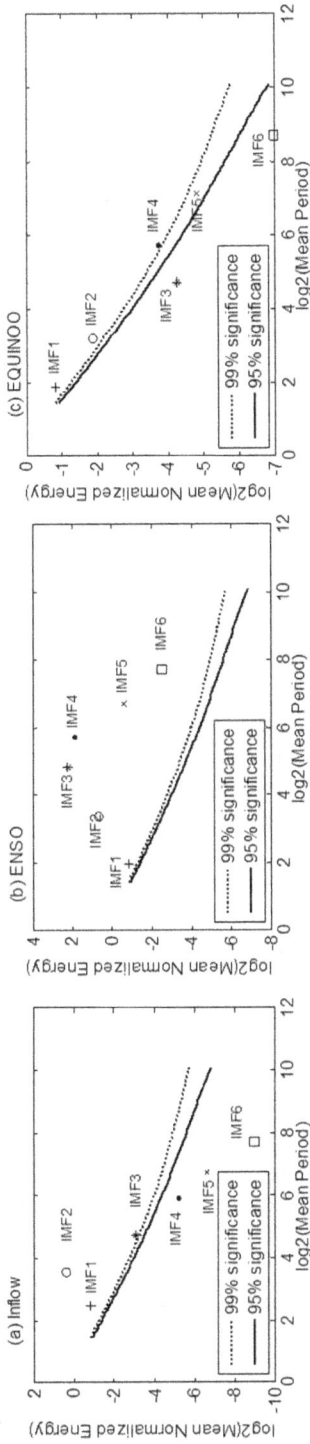

FIGURE 5.20 Statistical significance test of IMF components of (a) inflow into Hirakud reservoir; (b) ENSO and (c) EQUINOO. (From Adarsh 2017).

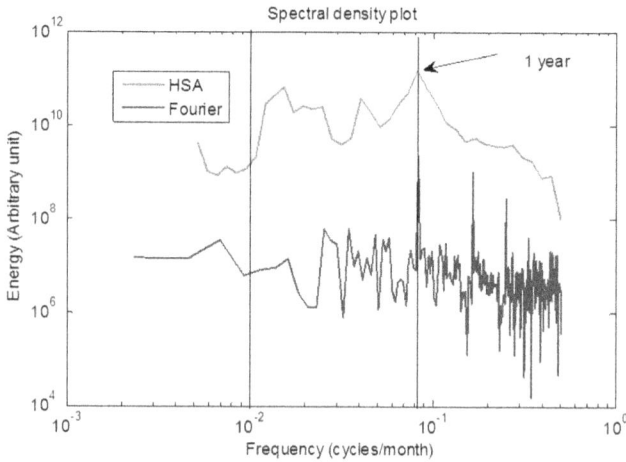

FIGURE 5.21 Marginal Hilbert spectrum and Fourier spectrum of inflows into Hirakud reservoir. (From Adarsh 2017).

the modes of inflow with that of ENSO/EQUINOO is performed and the results are presented in the Table 5.4. The cross correlation analysis between different modes shows that high correlation is noticed only between the residues of the EQUINOO and inflow (−0.98) followed by the residue of ENSO with that of inflow (−0.58). However, significant correlation between the modes is noticed for few cases, as highlighted in Table 5.4. From the table, it is clear that the nature of correlation between different modes is not of same character (values falling along the diagonal), i.e., at different time scales the association between the climate oscillation and inflow is not same. For example, it can be seen that a correlation of 0.343 for IMF3 and −0.214 for IMF4, when the association between ENSO and inflow is considered (i.e., at some of the time scales the association is positive while in some other time scales the association can be negative). On considering the association between EQUINOO and inflow, the correlation is −0.912 for IMF6, while that is 0.276 at IMF2. This is a remarkable observation. Furthermore, the linear correlation between inflow and the ENSO series is only −0.024, and that between inflow and EQUINOO series is 0.115. The overall correlation is found to be very low, which may be because of the reason that the positive correlation at some time scale may get nullified by the negative correlation in some other time scale.

To get a better insight to the strength of association and for capturing such alterations in a better way, a running correlation analysis can be very helpful. Therefore, the TDIC analysis (Chen et al. 2010a) is performed between the modes of inflow and that of oscillation (having comparable periodicity). The results of TDIC analysis of ENSO are presented in Figure 5.22 and that for EQUINOO are presented in Figure 5.23.

The long-range association between ENSO and inflow is noticed in IMF2-IMF5, i.e., the correlation is significant for time spells (window sizes) ranging from mean period of the respective mode to half of the data length. There is a dominancy of

TABLE 5.4
Correlation Between the Different Modes of Inflow Time Series with that of ENSO and EQUINOO Time Series

ENSO	Streamflow						
	IMF1	**IMF2**	**IMF3**	**IMF4**	**IMF5**	**IMF6**	**Residue**
IMF1	0.017	0.061	−0.001	−0.014	0.020	−0.025	−0.022
IMF2	0.019	−0.152	0.009	0.006	−0.016	0.015	0.016
IMF3	0.039	0.080	0.343	0.014	0.024	−0.017	−0.013
IMF4	−0.034	−0.042	0.006	−0.214	−0.144	0.072	0.022
IMF5	0.035	−0.034	−0.018	−0.158	0.228	−0.096	0.082
IMF6	0.017	0.028	−0.096	−0.004	0.099	−0.001	0.170
Residue	0.126	0.067	−0.034	0.152	−0.174	**−0.418**	**−0.583**

EQUINOO	Streamflow						
IMF1	−0.200	−0.229	0.002	0.003	0.042	0.011	−0.028
IMF2	−0.110	0.276	0.019	0.007	−0.004	0.026	0.050
IMF3	0.022	−0.019	0.087	0.028	0.025	−0.002	−0.033
IMF4	0.012	0.052	−0.115	0.116	0.018	0.001	0.102
IMF5	−0.013	0.066	0.095	−0.183	0.000	−0.142	−0.392
IMF6	−0.010	0.010	0.004	0.011	0.129	**−0.912**	−0.013
Residue	0.011	−0.011	−0.047	−0.151	0.128	0.054	**−0.984**

Note: The bold numbers indicate statistical significance at 5% significance level.

positive relation between the two in IMF3 and IMF5, while for the low frequency mode IMF6, it shows strong negative correlation. However, many switchovers in the nature of correlation can be noticed between the two series. For example, there is positive correlation between ENSO and reservoir inflow in 1968–1978 in IMF3, and then turns negative for a decade and then becomes negative. In IMF4, a dominancy of negative association is noticed, but it is positive during 1968–1973. Long range negative correlation is observed between different modes of EQUINOO with reservoir inflow, but in IMF2 and IMF3, the association is positive. A negative correlation between EQUINOO and inflow indicates the out-of-phase relationship between them. The reason for such negative association could be that the convective activity increases over the western part of Indian Ocean (Maity and Nagesh Kumar 2009). Being located in the eastern part of India, the rainfall might be lower-than-normal over the Mahanadi basin, which eventually cause in decreased inflow into the Hirakud reservoir. Also, it is noticed that the relation between the EQUINOO and reservoir inflow is not of unique character always. For example, a strong negative correlation between the two is noticed during 1965–1975 in IMF3, eventhough the association is primarily positive at this time scale. Similarly, the association between the two is primarily negative at the time scale of IMF4 while it is positive during 1973–1978. Such switchovers in the nature of association are relatively less in low frequency modes.

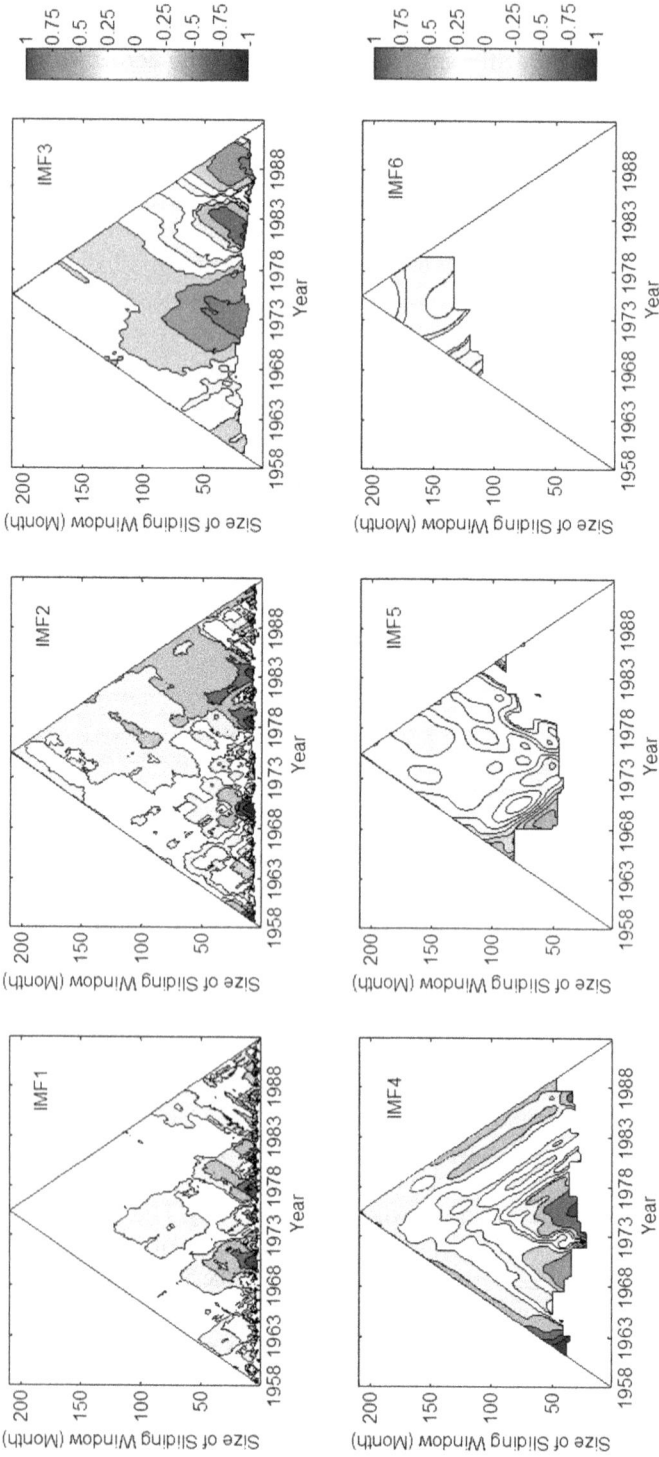

FIGURE 5.22 TDIC plots between the IMFs of ENSO and reservoir inflow time series. The white space of the TDIC plot means that correlation coefficient is not significant at 5% significance level.

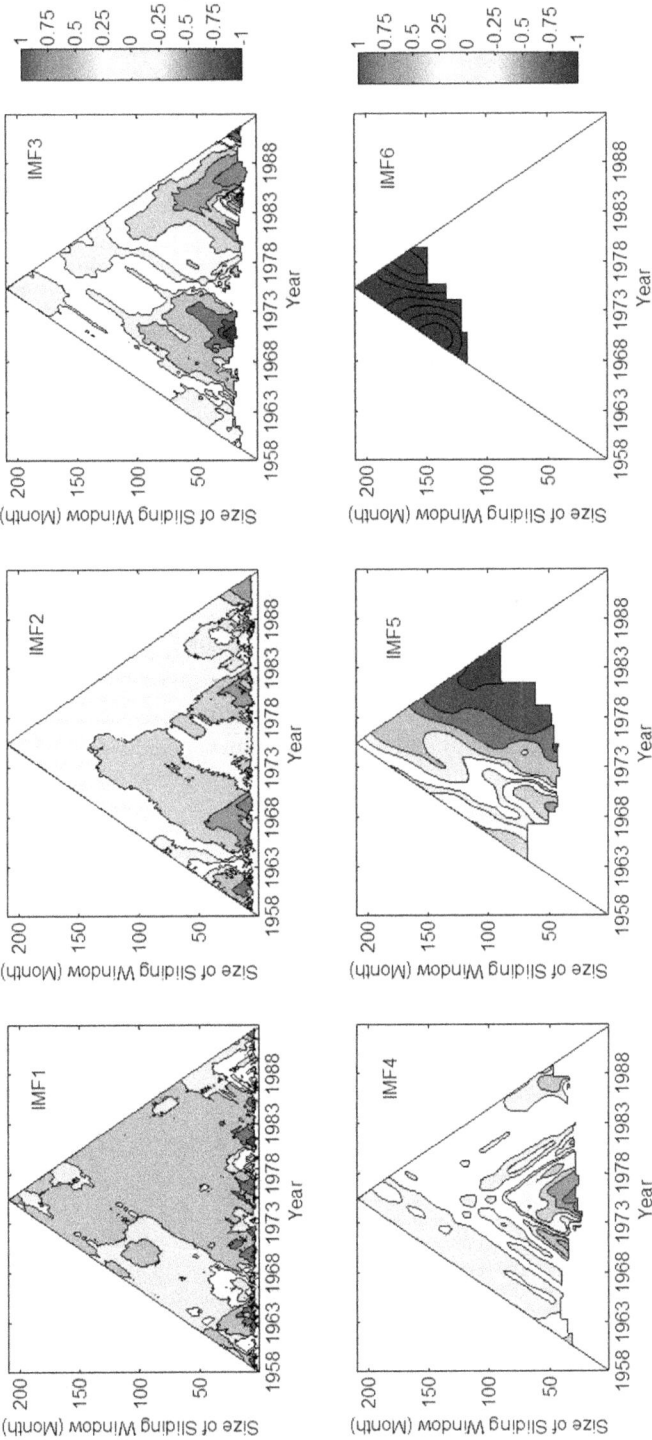

FIGURE 5.23 TDIC plots between the IMFs of EQUINOO and reservoir inflow time series. The white space of the TDIC plot means that correlation coefficient is not significant at 5% significance level.

In other words, the correlation pattern is fairly stable at lager time scales. In general, many such alterations in the nature of correlations are observed between the inflow and oscillations, particularly at high frequency modes.

5.4 CLOSURE

This chapter presented diverse practical applications of HHT on hydrologic datasets, streamflows in particular. The application of TDIC analysis has brought new insights and proven as an alternative practice for the multiscale investigations on streamflow-sediment linkages. Hence the case study applications of datasets of Basantpur and Tikrapara stations proven that TDIC can be used as alternative to wavelet coherence analysis for capturing the climatic or anthropogenic impacts of hydrological variability of river flows in India. From the applications, it is well evident that the application of this advanced spectral analysis procedure is not limited only to hydroclimatic teleconnections, but it is generic one and possible for investigating any of the hydrologic linkages. The application of AOHSA method has established that the daily time series of TSS concentration data of Mahanadi river are multifractal in character and the method is successful in capturing the scaling properties hydrological datasets of the river. This opens the scope for applications of AOHSA as an alternative to many of the existing formalisms of multifractal analysis. The concurrent roles of multiple climatic oscillations may be the reasons for streamflow variability. The MEMD based TDIC analysis proposed in this study could recognize the common scales inherent in different causal time series and streamflow, which is established as a powerful tool for hydro- climatic teleconnection studies in multiple time scales.

REFERENCES

Adarsh, S. 2017. *Multiscale characterization of hydrologic time series using mathematical transforms*. Ph.D. thesis, IIT Bombay, India.
Antico, A., Schlotthauer, G., Torres, M.E. 2014. Analysis of hydro-climatic variability and trends using a novel empirical mode decomposition: *Application to Parana River Basin. Journal of Geophysical Research: Atmospheres* **119**(3): 1219–1233.
Barnhart B.L., Eichinger W.E. 2011. Empirical mode decomposition applied to solar irradiance, global temperature, sunspot number and CO_2 concentration data. *Journal of Atmospheric and Solar-Terrestrial Physics* **73**(2011): 1771–1779.
Cao Z., Li Y., Yue Z. 2007. Multiple time scales of alluvial rivers carrying suspended sediment and their implications for mathematical modeling. *Advances in Water Resources* **30**: 715–729.
Chen, X., Wu, Z., Huang, N.E. 2010a. The time-dependent intrinsic correlation based on the empirical mode decomposition. *Advances in Adaptive Data Analysis* **2**: 233–265.
Chen, Y.D., Zhang, Q, Xu, C-Y., Lu, X., Zhang, S. 2010b. Multiscale streamflow variations of the Pearl river basin and possible implications for the water resource management within the Pearl river delta, China. *Quaternary International* **226**(2010): 44–53.
Flandrin, P., Gonçalvès, P. 2004. Empirical mode decompositions as data-driven wavelet-like expansions. *International Journal of Wavelets, Multi-resolution Information Processing* **2**(4): 477–496.

Flandrin, P., Rilling, G., Gonçalvés, P. 2004. Empirical mode decomposition as a filter bank. *IEEESignal Processing Letters* **11**(2): 112–114.

Gadgil S., Vinayachandran P.N., Francis P.A., Gadgil S. 2004. Extremes of the Indian summer monsoon rainfall, ENSO and equatorial Indian Ocean oscillation. *Geophysical Research Letters* 31 L12213 doi:10.1029/2004GL019733.

Gray, A.B., Pasternack, G.B., Watson, E.B., Warrick, J.A., Goni, M.A. 2014. The effect of El Niño southern oscillation cycles on the decadal scale suspended sediment behavior of a coastal dry-summer subtropical catchment. *Earth Surface Processes and Landforms* **40**(2): 272–274.

Hu W., Si B-C. 2013. Soil water prediction based on its scale-specific control using multivariate empirical mode decomposition. *Geoderma* **193-194**: 180–188.

Huang, Y.X. 2009. *Arbitrary order Hilbert spectral analysis: definition and application to fully developed turbulence and environmental time series.* Ph.D. Thesis in Fluid Dynamics. University of Lille, France.

Huang, Y., Schmitt, F.G, Lu, Z., Liu, Y. 2009a. Analysis of daily river flow fluctuations using empirical mode decomposition and arbitrary order Hilbert spectral analysis. *Journal of Hydrology* **373**: 103–111.

Huang, N.E, Wu, Z., Long, S.R., Arnold, K.C., Blank, K., Liu, T.W. 2009b. On instantaneous frequency. *Advances in Adaptive Data Analysis* **1**(2): 177–229.

Kantelhardt, J.W., Koscielny-Bunde, E., Rybski, D., Braun, P., Bunde, A., Havlin, S. 2006. Long-term persistence and multifractality of precipitation and river runoff records. *Journal of Geophysical Research* **111**:D01106. doi:10.1029/2005JD005881.

Kuai, K.Z., Tsai, C.W. 2012. Identification of varying time scales in sediment transport using the Hilbert–Huang Transform method. *Journal of Hydrology* **420–421**(2012): 245–254.

Li, E., Mu, X., Zhao, G., Gao, P. 2015. Multifractal detrended fluctuation analysis of streamflow in the Yellow river basin, China. *Water* **7**: 1670–1686.

Lombardo, F., Volpi, E., Koutsoyiannis, D., Papalexiou, S. 2014. Just two moments! A cautionary note against use of high-order moments in multifractal models in hydrology. *Hydrology and Earth System Sciences* **18**: 243–255.

Maity, R., Nagesh Kumar, D. 2006a. Bayesian dynamic modeling for monthly Indian summer monsoon rainfall using El Niño-Southern Oscillation (ENSO) and Equatorial Indian Ocean Oscillation (EQUINOO). *Journal of Geophysical Research – Atmospheres* **111**, D07104, doi:10.1029/2005JD006539.

Maity, R., Nagesh Kumar, D. 2006b. Hydroclimatic association of monthly summer monsoon rainfall over India with large-scale atmospheric circulation from tropical Pacific Ocean and Indian Ocean region. *Atmospheric Science Letters* **7**(4): 101–107.

Maity, R., Nagesh Kumar, D. 2007a. Hydroclimatic teleconnection between global sea surface temperature and rainfall over India at subdivisional monthly scale. *Hydrological Processes* **21**(14): 1802–1813.

Maity, R., Nagesh Kumar, D., Nanjundiah, R.S. 2007b. Review of Hydroclimatic teleconnection between hydrologic variables and large-scale atmospheric circulation patterns with Indian perspective. *ISH Journal of Hydraulic Engineering* **13**(1): 77–92.

Maity, R., Nagesh Kumar, D. 2008. Basin-scale streamflow forecasting using the information of large-scale atmospheric circulation phenomena. *Hydrological Processes* **22**(5): 643–650.

Maity, R., Nagesh Kumar, D. 2009. Hydroclimatic influence of large-scale circulation on the variability of reservoir inflow. *Hydrological Processes* **23**(6): 934 –942.

Milliman, J.D., Syvitski, J.P.M. 1992. Geomorphic/tectonic control of sediment discharge to the ocean: The importance of small mountainous rivers. *The Journal of Geology* **100**: 5425–5441.

Pandey, G., Lovejoy, S., Schertzer, D. 1998. Multifractal analysis of daily river flows including extremes for basins five to two million square kilometers, one day to 75 years. *Journal of Hydrology* **208**: 62–81.

Sadeghi, S.H.R., Mostafazadeh, R. 2016. Triple diagram models for changeability evaluation of precipitation and flow discharge for suspended sediment load in different time scales. *Environmental Earth Science* **75**(843) DOI: 10.1007/s12665-016-5621-6.

Shang, P., Kamae, S. 2005. Fractal nature of time series in the sediment transport phenomenon. *Chaos, Solitons Fractals* **26**: 997–1007.

Thompson, S.E., Katul, G.G. 2012. Multiple mechanisms generate Lorentzian and 1/f α power spectra in daily stream-flow time series. *Advances in Water Resources* **37**: 94–103.

Torres, M.E., Colominas, M.A., Schlotthauer, G., Flandrin, P. 2011. A complete ensemble empirical mode decomposition with adaptive noise. *IEEE International conference on Acoustic Speech and Signal Processing*, Prague 22–27 May 2011, pp. 4144–4147.

Walling, D.E., Fang, D. 2003. Recent trends in the suspended sediment loads of the world's rivers *Global and Planetary Change* **39**(2003): 111–126.

Wu, Z., Huang, N.E. 2004. Statistical significance test of intrinsic mode functions. In *Hilbert-Huang Transform and its Applications*. Edited by: Norden E. Huang (NASA Goddard Space Flight Center, USA), Samuel S.P. Shen (University of Alberta, Canada). Singapore: World Scientific Publishing. pp. 149–169.

Zhang, Q., Chen, G., Su, B., Disse, M., Jiang, T., Xu, C-Y. 2008. Periodicity of sediment load and runoff in the Yangtze river basin and possible impacts of climatic changes and human activities. *Hydrological Sciences Journal* **53**(2): 457–464.

Zhang, Q., Singh, V.P., Li, K., Li, J. 2014. Trend, periodicity and abrupt change in streamflow of the East River, the Pearl River basin. *Hydrological Processes* **28**(2): 305–314.

Zhang, Q., Singh, V.P., Xu, C., Chen, X. 2013. Abrupt behaviors of streamflow and sediment load variations of the Yangtze river basin, China. *Hydrological Processes* **27**: 444–452.

Zhang, Q., Xu, C., Chen, X., Lu, X. 2012. Abrupt changes in the discharge and sediment load of the Pearl River, China. *Hydrological Processes* **26**: 1495–1508.

Zhang, Q., Xu, C-Y., Singh, V.P, Yang, T. 2009. Multiscale variability of sediment load and streamflow of the lower Yangtze river basin: Possible causes and implications. *Journal of Hydrology* **368**: 96–104.

6 MEMD-based Hybrid Schemes for Hydrological Modeling

6.1 BACKGROUND

The characterization of hydrological time series and identification of dominant predictor variables can be done efficiently in multiple time scales and in the time-frequency domain, with the help of advanced spectral analysis methods such as wavelet transforms and HHT. Such exercise will have more practical appeal, only when the information gathered is extended to real field applications such as simulation or prediction problems. As multiple variables are responsible for the hydrological processes like rainfall, streamflow or drought, the advantage of MEMD can be utilized effectively for simulation problems in hydrology. This brings the advantages of equal number of modes for different variables on decomposition and gathering useful information from multiple process scales. The usefulness of MEMD lies primarily as a preprocessing decomposition tool in the simulation or prediction problems. Subsequently, the modes obtained can be simulated individually using linear regression method or nonlinear data-driven paradigms, considering the appropriate inputs at different process scales. Final step of aggregation of estimated modes at different scales enable to get the hydrological variable of concern. This hybrid MEMD framework is generic and described in detail in Chapter 2. The application of this proposed framework is demonstrated considering different datasets like rainfall, drought index, inflow and suspended sediment of different temporal resolutions varying from annual to daily scales.

6.2 PREDICTION OF SEASONAL RAINFALL USING MEMD-SLR HYBRID MODEL

The monthly rainfall data of Kerala meteorological subdivision (8.5°N, 76.98°E) in India is selected for demonstration of the proposed approach of MEMD-SLR hybrid model for seasonal predictions. The State of Kerala, located in the western coast of Southern India, receives a high average annual rainfall (~300 cm) when compared with that of the country (~120 cm). Here the summer monsoon season (June–July–August–September, JJAS) is the prominent rainfall season in which the state receives more than 60% of the annual rainfall. The monthly summer monsoon rainfall data for the period 1950–2012 are collected from IITM Pune, and the data of four different climate oscillations are collected from different organizations, as described in

Chapter 4. From the investigation of hydroclimatic teleconnections of seasonal rainfall in Kerala, it is well established that the rainfall series is modulated by different climatic oscillations such as QBO, ENSO, EQUINOO and AMO (Adarsh and Janga Reddy 2016). The rainfall data and the data of four climatic oscillations together constitute the multivariate dataset. To apply MEMD methodology, the number of direction vectors is selected as 64, and the upper and lower bound of threshold parameters are chosen as 0.075, 0.75 and the splitting parameter is chosen as 0.075, following the suggestions made in the past works (Rilling et al. 2003; Hu and Si 2013), and performed the decomposition of multivariate dataset. MEMD resulted in seven IMFs and residue for all the time series. The mean periods of the different modes computed by zero-crossing method (Barnhart and Eichinger 2011) are given in Table 6.1. The mean period presented in Table 6.1 also confirm the association of interannual oscillations like QBO (2- to 3-year periodicity), ENSO (of 7- to 8-year periodicity), EQUINOO (of interannual periodicity) and AMO (interdecadal periodicity).

Among the different climatic oscillations, the concurrent role of ENSO and EQUINOO on ISMR is well debated in literature (Gadgil et al. 2004; Maity and Nagesh Kumar 2006a, 2006b), and multiple past lagged values of these indices may be influential to the rainfall of a generic month and hence that of the season. First the monsoon rainfall, the data of four monsoon months (JJAS) are considered. The previous five years lagged value of rainfall of the particular month, ENSO and EQUINOO indices of three previous months are considered as input following the feedbacks from many previous studies (Sahai et al. 2000; Kashid and Maity 2012; Singh and Borah 2013). For example, for prediction of August rainfall, previous five-year June rainfall values, ENSO and EQUINOO indices of March, April and May are used as input. Hence total 11 inputs will be there for prediction of rainfall value of each month. The model can have a functional relationship in the form of:

TABLE 6.1
Mean Period (In Years) of Rainfall and Indices of Climate Oscillations and the Average of Process Scales

Mode Number	Mean Period					Average Period
	Rainfall	QBO	ENSO	EQUINOO	AMO	
IMF1	1.1	1.0	1.1	1.0	1.1	1.0
IMF2	2.2	2.5	2.4	2.2	2.4	2.2
IMF3	4.0	3.4	4.2	3.7	4.0	3.7
IMF4	6.5	6.5	7.6	7.0	7.6	7.6
IMF5	14.0	14.0	14.0	12.0	10.5	16.8
IMF6	28.0	21.0	28.0	21.0	28.0	28.0
IMF7	84.0	84.0	84.0	84.0	42.0	42.0
Residue	LT	84.0	LT	84.0	LT	LT

Note: LT refers to long term.

$$R_{tJune} = f(R_{June(t-1)} , ... R_{June(t-5)}, EN_{tMay},$$
$$EN_{tApril}, EN_{tMarch}, EQ_{tMay}, EQ_{tApril}, EQ_{tMarch})$$

Then the seasonal rainfall (monsoon season) can be calculated as

$$R_{monsoon} = R_{June} + R_{July} + R_{August} + R_{September}$$

For modeling rainfall of month June, the multivariate dataset comprising 12 input parameters is first decomposed by the MEMD method. The threshold parameters are selected as 0.075, 0.75 respectively, the splitting parameter is chosen as 0.075 and the number of direction vectors is selected as 64. The decomposition resulted in five IMFs and residue. The stepwise linear regression (SLR) models are developed for individual modes of rainfall. The regression coefficients, which are found to be insignificant (based on p-value statistics) are brought to zero. First 41 years of data are used for model calibration and the rest of the data are used for validation. The regression coefficient matrix for rainfall prediction of June month is given in Table 6.2.

Hence the equation for prediction of IMF1 of June rainfall will be

$$IMF1_{R_{tJune}} = -1.734 IMF1_{R_{t-1}} - 1.827 IMF1_{R_{t-2}} - 1.326 IMF1_{R_{t-3}} + 1.978$$

TABLE 6.2
Regression Coefficient Matrix of Models Fitted to Predict the Rotational Modes of June Rainfall In Kerala Subdivision.

Modes of	Modes of rainfall					
	IMF1	IMF2	IMF3	IMF4	IMF5	Residue
R_{t-1}	−1.734	0.424	0.679	2.253	2.109	1.148
R_{t-2}	−1.827	−1.236	−0.034	−2.122	−2.078	0.000
R_{t-3}	−1.326	0.365	0.031	−0.150	0.202	0.683
R_{t-4}	−0.383	−1.226	0.003	1.687	1.134	0.000
R_{t-5}	−0.274	0.029	−0.514	−1.152	−0.601	−0.619
EN_{May}	13.793	43.529	10.566	−5.817	−134.508	0.000
EN_{April}	29.484	198.957	−22.393	−102.853	−26.100	0.000
EN_{March}	35.282	−170.273	−13.428	218.699	252.467	−132.982
EQ_{May}	−22.941	31.526	25.039	−27.924	−43.747	11.929
EQ_{April}	14.758	47.121	12.874	32.250	45.362	0.858
EQ_{March}	6.369	68.790	−36.301	−78.526	−80.822	0.000
Constant	1.978	0.895	−0.257	0.122	−1.483	−168.173

Note: The numbers in italics indicate the p-value is not acceptable and the coefficients are to be brought to zero in the models of different subseries.

Similarly, the equation for prediction of IMF3 of June rainfall will be

$$IMF3_{R_{t,June}} = 0.679 IMF3_{R_{t-1}} - 0.514 IMF3_{R_{t-5}} + 25.039 IMF3_{EQ_{May}}$$
$$- 36.3 IMF3_{EQ_{March}} - 0.257$$

In a similar way, the equations for prediction of other modes can be deduced. Final summation of the predicted modes provides the rainfall of June month. Based on the models developed, the modes of validation data are predicted, and the summation of these predicted modes provide the predicted values rainfall for validation period. Similar procedure is followed to predict the rainfall of July, August, September months also and the regression coefficients of modes for prediction of the rainfall in these months are provided in Tables 6.3–6.5 The decomposition resulted in six modes for the month of July, seven modes for August and eight modes for September.

Time series plot of observed and predicted rainfall of calibration period and validation period for four monsoon months are provided in Figure 6.1.

The different performance evaluation measures for the calibration and validation data for the four months are summarized in Table 6.6. Different statistical evaluation measures such as correlation coefficient (R). Nash-Sutcliff Efficiency (NSE), Index of Agreement (IA), Root Mean Square Error (RMSE), Mean Absolute Error (MAE) and Mean Bias Error (MBE) are used to evaluate the potential of the proposed approach.

TABLE 6.3
Regression coefficient matrix of models fitted to predict the rotational modes of July rainfall

Modes of	Modes of rainfall					
	IMF1	IMF2	IMF3	IMF4	IMF5	Residue
R_{t-1}	−2.234	0.291	1.729	2.745	2.425	1.097
R_{t-2}	−3.348	−1.056	−2.258	−2.978	−2.125	0.000
R_{t-3}	−3.327	*0.170*	1.332	1.277	0.120	*−0.035*
R_{t-4}	−2.233	−0.627	−0.667	*0.075*	1.400	*−0.018*
R_{t-5}	−0.933	*−0.118*	*0.032*	−0.237	−0.858	*0.002*
EN_{June}	*−10.764*	12.568	69.158	104.250	19.408	−67.232
EN_{May}	−46.356	−23.153	−145.498	5.615	−0.612	153.195
EN_{April}	56.017	−47.699	118.572	−133.503	−44.465	10.382
EQ_{June}	*22.144*	−17.350	−28.963	*0.508*	35.536	−20.621
EQ_{May}	−0.597	6.463	−25.045	*0.083*	*0.185*	−5.695
EQ_{April}	*5.416*	−30.448	*8.843*	13.207	−9.980	−11.920
Constant	1.3049	2.7593	0.1488	−0.6268	0.0012	−25.5055

Note: The numbers in italics indicate the *p*-value is not acceptable and the coefficients are to be brought to zero in the models of different subseries.

TABLE 6.4
Regression Coefficient Matrix of Models Fitted to Predict the Rotational Modes of August Rainfall

Modes of	Modes of rainfall						
	IMF1	IMF2	IMF3	IMF4	IMF5	IMF6	Residue
R_{t-1}	−2.051	*−0.148*	0.919	1.419	*−0.005*	−0.795	1.599
R_{t-2}	−2.632	−1.165	−0.355	*0.138*	*0.016*	−0.769	−0.434
R_{t-3}	−2.304	*−0.221*	−0.504	−0.592	2.014	5.018	0.000
R_{t-4}	−1.478	−0.428	0.282	*−0.042*	−1.202	−1.530	−0.359
R_{t-5}	*−0.548*	*0.047*	*−0.019*	0.164	*−0.184*	−1.295	0.018
EN_{July}	*−10.974*	*3.354*	63.684	*−10.312*	1.486	−482.264	0.000
EN_{June}	−2.936	6.612	7.276	2.977	−32.552	417.478	0.000
EN_{May}	−24.960	*13.742*	7.595	24.429	*4.013*	−66.904	0.000
EQ_{July}	−7.798	−22.617	−16.371	30.932	24.571	−47.036	5.706
EQ_{June}	*1.228*	−5.906	*−1.478*	−5.275	*0.254*	67.969	−3.123
EQ_{May}	5.499	*−1.912*	−23.578	*−0.014*	*0.328*	−11.531	0.000
Constant	2.706	0.369	0.658	−0.062	−0.062	0.000	71.119

Note: The numbers in italics indicate the p-value is not acceptable and the coefficients are to be brought to zero in the models of different subseries.

TABLE 6.5
Regression Coefficient Matrix of Models Fitted to Predict the Rotational Modes of September Rainfall

Modes of	Modes of rainfall							
	IMF1	IMF2	IMF3	IMF4	IMF5	IMF6	IMF7	Residue
$Rt-1$	−2.541	0.369	0.910	0.773	*0.074*	0.603	1.580	0.000
$Rt-2$	−3.611	−1.093	−1.236	*0.163*	*0.108*	1.685	0.000	1.170
$Rt-3$	−3.558	*−0.119*	0.936	0.855	*0.036*	1.462	0.000	0.000
$Rt-4$	−2.266	−0.326	−0.767	−2.047	*0.012*	−2.281	−0.671	0.000
$Rt-5$	−0.786	−0.020	*0.142*	1.169	0.025	1.663	1.741	0.000
EN_{August}	*0.315*	−22.324	205.053	78.621	*2.165*	77.443	−85.599	−204.212
EN_{July}	*−0.494*	−14.414	−363.679	−57.779	84.536	−102.320	82.839	245.958
EN_{June}	*0.647*	−21.165	80.676	*12.124*	*−13.409*	29.739	−97.401	7.212
EQ_{August}	25.043	2.399	−3.265	−47.288	*1.184*	−34.456	−109.105	0.000
EQ_{July}	*−3.045*	−18.397	−28.531	15.056	39.082	1.513	−5.291	10.298
EQ_{June}	−10.179	*1.281*	*0.409*	32.661	2.439	*−9.883*	85.332	0.000
Constant	−0.021	0.108	0.127	−0.124	0.117	−0.001	0.000	−34.684

Note: The numbers in italics indicate the p-value is not acceptable and the coefficients are to be brought to zero in the models of different subseries.

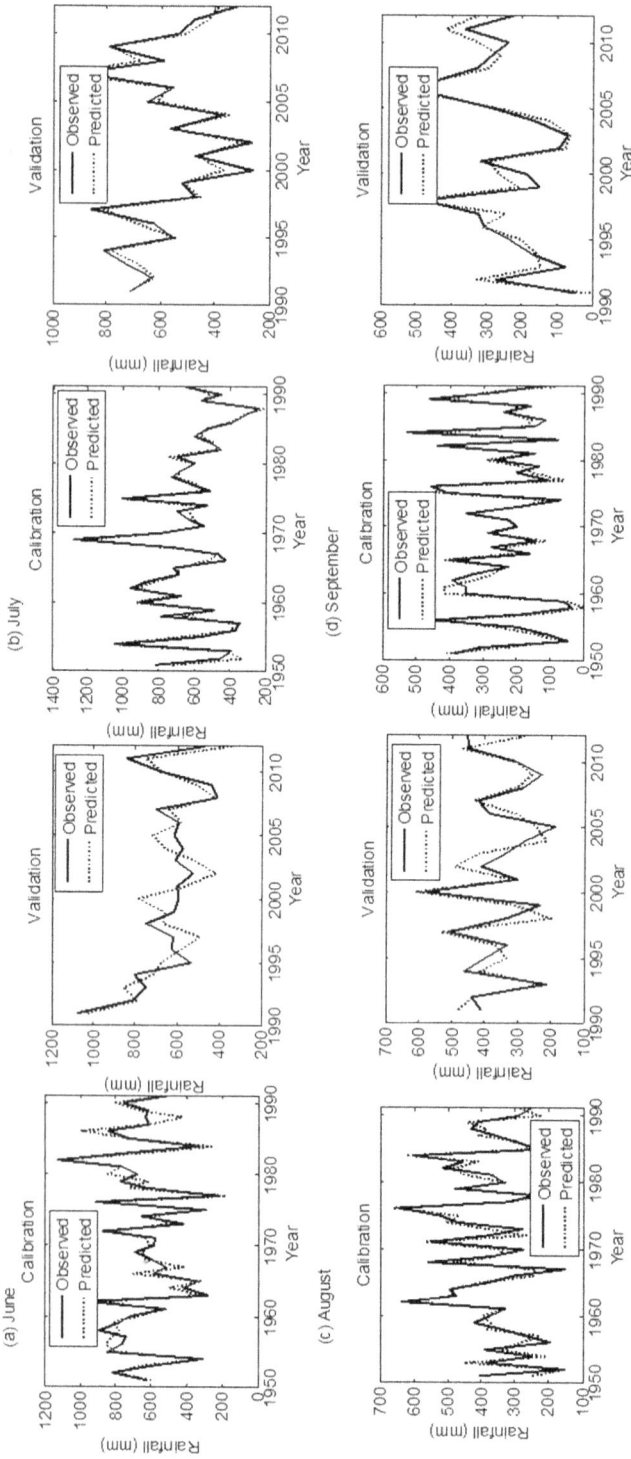

FIGURE 6.1 Time series plots of rainfall predictions for different monsoon months (JJAS) for calibration period and validation period.

TABLE 6.6
Performance Evaluation of Rainfall Predictions During Monsoon Months (JJAS) by MEMD-SLR Method

PM	June		July		August		September	
	Cal	Val	Cal	Val	Cal	Val	Cal	Val
R	0.927	0.824	0.974	0.971	0.903	0.894	0.959	0.930
NSE	0.850	0.626	0.947	0.942	0.808	0.639	0.920	0.839
IA	0.808	0.663	0.889	0.883	0.788	0.711	0.858	0.805
RMSE	77.51	89.313	48.720	42.25	53.41	67.73	37.55	48.61
MAE	61.156	76.772	35.731	32.20	23.0	52.00	31.73	39.78
MBE	−1.477	9.843	−0.106	−4.60	1.662	0.351	−0.400	−9.60

Note: Cal – calibration; Val – validation; PEC is the performance evaluation criteria.

From different performance evaluation measures presented in Table 6.6, it is clear that in all monsoon months, the predictions using MEMD-SLR method are found to be reasonably well with the largest error measures for the predictions of June month. This is obvious because eventhough summer monsoon commences in the month of June, the monsoon dynamics might not have fully developed in this month. Then the high rainfall values (rainfall exceeding mean + SD) and low rainfall (rainfall below mean − SD) are segregated and rainfall predictions for these cases were compared with observed data of respective series (of different months) in Figure 6.2. On quantifying the error in predictions for validation data in terms of RMSE, it is noticed that the RMSE for prediction of high rainfall are 70.03, 34.48, 26.65 and 15.91, respectively, for June, July, August and September months, where the RMSE statistics for prediction of low rainfall for these months are 84.62, 65.13, 50.92 and 42.53. Hence it can be inferred that the MEMD-SLR model is better in capturing high rainfall than the low rainfall. The addition of predicted rainfalls of June, July, August and September months gives the rainfall for monsoon season.

A similar modeling exercise is performed upon the postmonsoon rainfall of the subdivision by considering the five previous year values and three most correlated monthly series of ENSO and EQUINOO as inputs. The decomposition resulted in six modes for October rainfall and seven modes for November rainfall. The regression coefficients of the modes of rainfall for October and November are provided in Table 6.7 and Table 6.8, respectively.

The addition of predicted rainfalls of October and November months gives the rainfall for post-monsoon season. The time series plots of monsoon and postmonsoon rainfall (observed and predicted data) of validation period (1991–2012) are presented in Figure 6.3.

The performance evaluation measures and the statistical properties of prediction were computed and presented in Table 6.9.

The performance evaluation statistics given in Table 6.9 show that the RMSE is less (136.09 and 72.96 for monsoon rainfall and post monsoon rainfall, respectively)

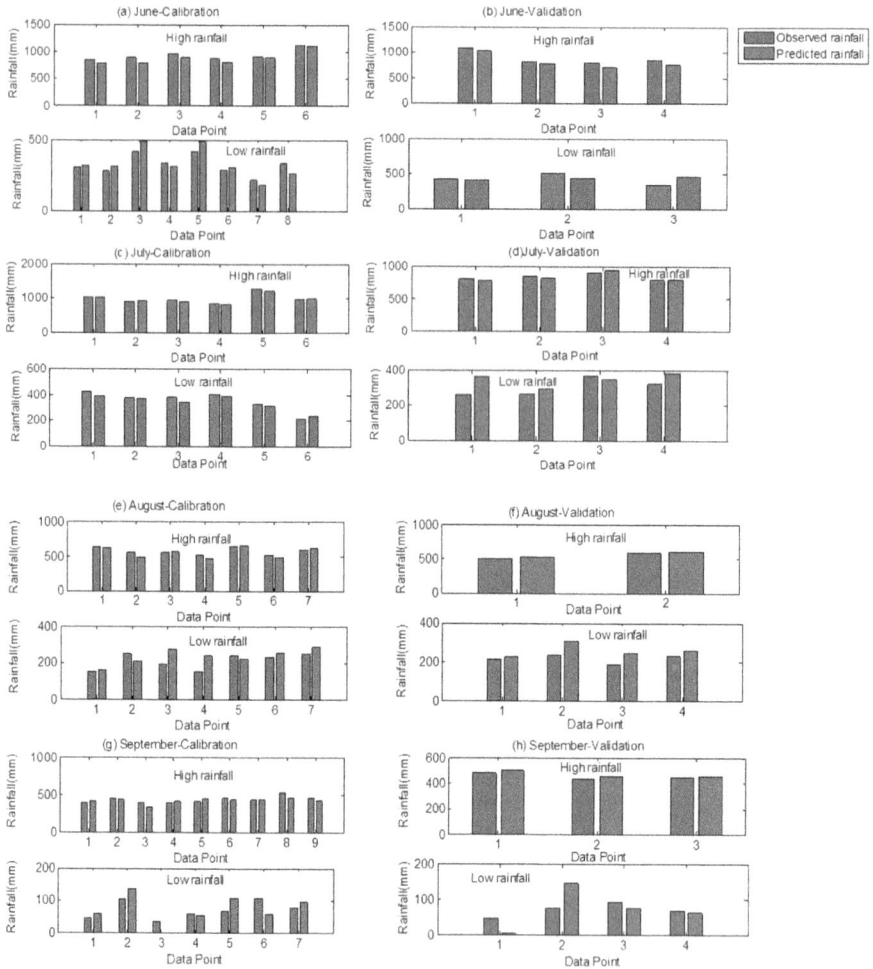

FIGURE 6.2 Observed and predicted high and low rainfall values for calibration and validation period.

and R value is reasonably good (0.90 and 0.866 for monsoon rainfall and post-monsoon rainfall respectively) for the predictions. The IA is also found to be satisfactory (0.796 and 0.741) in the complex problem of seasonal rainfall predictions at subdivisional scale in the Indian context (Kashid and Maity 2012). The percentage change in prediction of basic statistical properties maximum, minimum, mean and standard deviation are 0.47, 0.89, 0.22 and 7.6 (numerically) respectively for monsoon season. The percentage change in prediction of these properties is 15, 21, 1.89 and 9.1 (numerically) for postmonsoon season. Hence these statistics infer that the performance is satisfactory for seasonal rainfall predictions in Kerala with better accuracy for prediction of monsoon rainfall when compared to post-monsoon rainfall.

TABLE 6.7

Regression Coefficient Matrix of Models Fitted to Predict the Rotational Modes of October Rainfall Along with the Fitting Statistics

Modes of	Modes of rainfall					
	IMF1	IMF2	IMF3	IMF4	IMF5	Residue
R_{t-1}	−2.333	*0.028*	0.829	1.878	1.570	4.392
R_{t-2}	−3.053	−1.138	−0.279	−1.344	−2.761	−4.664
R_{t-3}	−2.593	*−0.045*	*0.105*	0.486	3.131	−0.551
R_{t-4}	−1.351	*−0.343*	*0.084*	*−0.037*	−1.172	1.709
R_{t-5}	*−0.345*	*−0.026*	−0.185	*−0.015*	−0.153	0.322
$EN_{September}$	*14.393*	*−11.199*	28.228	*2.291*	*−0.432*	−46.773
EN_{August}	*−5.591*	*−12.100*	*−11.546*	20.221	30.735	200.683
EN_{July}	*4.532*	38.075	−31.042	*−2.850*	*1.233*	−60.565
$EQ_{September}$	−6.225	−6.846	−19.697	*0.567*	−10.216	15.437
EQ_{August}	*−2.207*	8.007	−20.868	*−1.148*	−29.229	*−2.744*
EQ_{July}	*4.745*	*−11.182*	10.578	20.078	13.566	−6.096
Constant	−1.404	0.318	0.086	−0.003	0.053	−38.425

Note: The numbers in italics indicate the *p*-value is not acceptable and the coefficients are to be brought to zero in the models of different subseries.

TABLE 6.8

Regression Coefficient Matrix of Models Fitted to Predict the Rotational Modes of November Rainfall Along With the Fitting Statistics

Modes of	Modes of rainfall						
	IMF1	IMF2	IMF3	IMF4	IMF5	IMF6	Residue
Rt-1	−2.079	0.137	0.824	1.512	1.283	0.654	1.558
Rt-2	−2.524	−0.621	*0.128*	*−0.122*	−1.030	−1.527	0.000
Rt-3	−1.988	−0.485	−0.242	*−0.169*	4.313	5.233	0.000
Rt-4	−0.991	*−0.198*	*0.046*	−0.614	−5.152	−5.050	−0.433
Rt-5	−0.276	−0.086	*−0.019*	0.441	1.665	1.580	0.021
$EN_{October}$	4.851	4.171	−33.363	−53.327	*−12.129*	−427.122	0.000
$EN_{September}$	−7.103	−1.728	*14.656*	*34.593*	−8.990	444.714	0.000
EN_{August}	−9.314	6.481	27.341	96.553	−38.596	−22.567	−13.633
$EQ_{October}$	*1.760*	−10.940	*10.569*	99.662	−35.997	22.244	0.000
$EQ_{September}$	−4.438	4.251	46.265	−30.557	−0.528	−1.598	0.000
EQ_{August}	*1.340*	4.952	−2.710	*3.543*	8.500	−13.924	−26.631
Constant	0.689	−0.701	0.063	−0.188	−0.091	0.010	−40.492

Note: The numbers in italics indicate the *p*-value is not acceptable and the coefficients are to be brought to zero in the models of different subseries.

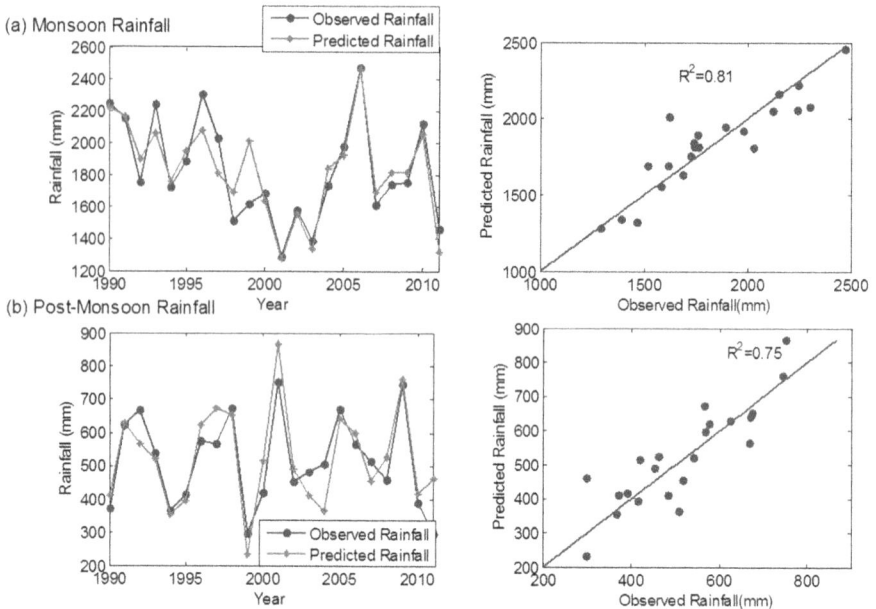

FIGURE 6.3   (a) Time series plots and (b) scatter plot of monsoon and post-monsoon rainfall prediction for the period 1991–2012.

6.3   MODELING SHORT-TERM DROUGHT USING THE MEMD-BASED HYBRID MODELS

Drought is one of the less attended but damaging natural calamities which occurs in many parts of the world annually. Indian economy is highly reliant on agricultural production which in turn heavily relies on natural rainfall. Any anomaly in the rainfall could lead to droughts and severe impact on the economy of our country, which is often characterized as meteorological drought (Mishra and Singh 2011; Zargar et al. 2011). Many studies have been focused on variability of droughts in different parts of India and for the whole India (Pai et al. 2011; Kumar et al. 2013; Ganguli and Reddy 2013; Thomas et al. 2015; Thomas and Prasannakumar 2016; Mallya et al. 2016). In most of the studies, the characterization of drought is made based on precipitation and the Standardized Precipitation Index (SPI) is the most widely used drought indicator owing to the flexibility of time scale, the stability of the spatial structure, the requirement of fewer input variables, and the simplicity of calculation (Bazrafshan et al. 2014). SPI of different time scales can indicate short-term as well as long-term droughts and has significant implications on different hydrologic components. The 3-month SPI can be used as a seasonal drought index to represent short-term drought and soil moisture may be more sensitive to a 3-month SPI representing the water stress and crop failures (Thomas et al. 2015). The prediction of droughts (SPI-3) is done following a pure time series approach or cause-effect approach using traditional time series or data driven techniques in which the lagged time step values of SPI or

TABLE 6.9
Performance Evaluation of Rainfall Prediction by MEMD-SLR Model Developed for Monsoon and Post-Monsoon Rainfall in Kerala (1991–2012)

Performance Measure	Monsoon Value	Post-Monsoon Value
R	0.901	0.866
NSE	0.811	0.694
IA	0.796	0.741
RMSE	136.092	72.962
MAE	100.564	58.689
MBE	−4.113	−9.769
SI	0.074	0.160
MXO	2470.700	751.5
MINO	1292.200	297.2
MXP	2459.010	865.77
MINP	1280.676	234.76
MNO	1832.777	516.10
SDO	320.808	134.956
MNP	1836.891	525.869
SDP	296.451	147.196

Note: MXO and MXP –maximum of observed data and maximum of predicted data; MINO and MINP – minimum of observed data and minimum of predicted data; MNO and MNP mean of observed data and mean of predicted data; SDO and SDP – standard deviation of observed data and standard deviation of predicted data.

the climatic indices have been used as inputs (Choubin et al. 2016). Recently wavelet analysis was used to develop hybrid models for prediction of SPI which improved the predictability efforts of SPI (Kim and Valdes 2003; Belayneh et al. 2014, 2016a, 2016b). The challenges on applying wavelets in terms of types and decomposition level along with the necessity of accounting the use of multiple causal inputs makes the modeling with MEMD potential alternatives for drought simulations.

Few studies reported that there is a decreasing trend of monsoon rainfall in Kerala (Krishnakumar et al. 2009; Adarsh and Janga Reddy 2015; Thomas and Prasannakumar 2016). In addition, Kerala was declared as drought affected region by the state government in every year since 2012. Along with this, two subdivisions Orissa and Telangana, which are climatically sensitive are considered in this study. The monthly rainfall data for the period 1871–2013 is collected from IITM Pune. The SPI-3 values of these three regions were estimated using the popular and standard procedure proposed by McKee et al. (1993) by using the two parameter Gamma distribution. For the selection of inputs, the Partial Auto Correlation (PAC) values were estimated. The SPI-3 series of the three subdivisions along with the PAC plots of SPI-3 series are presented in Figure 6.4.

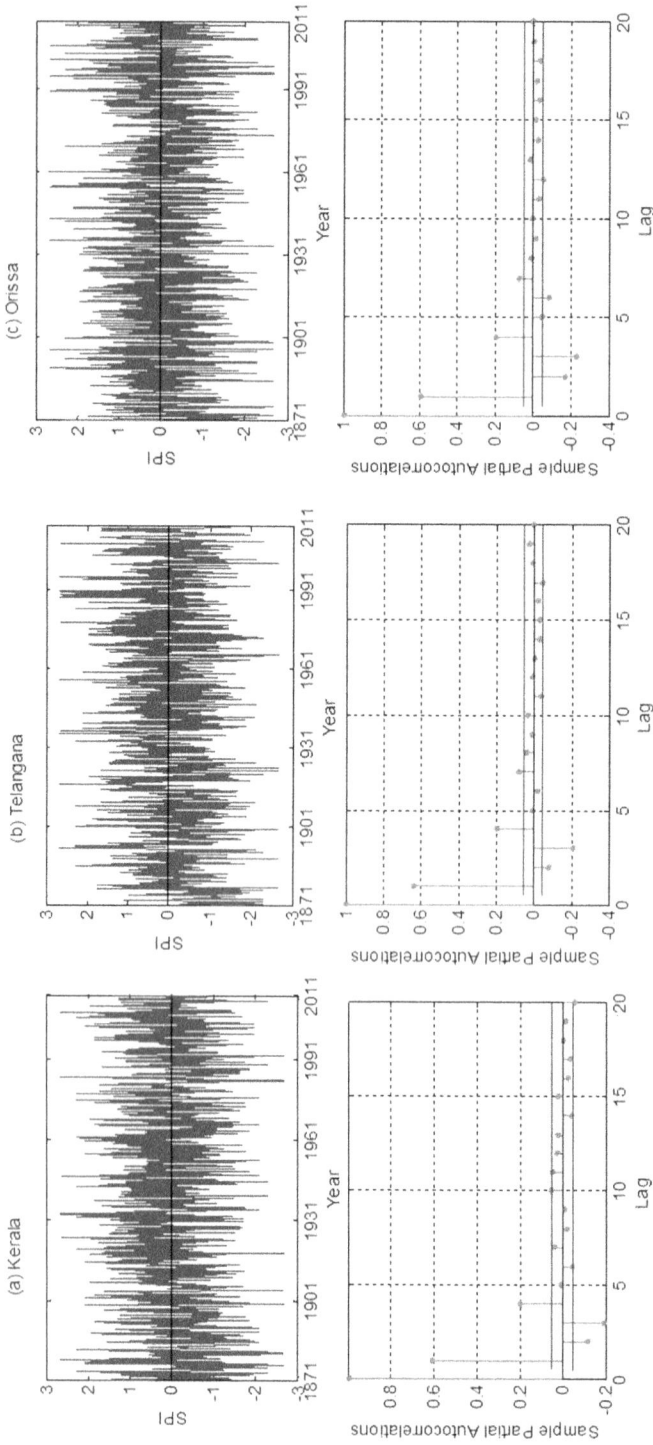

FIGURE 6.4 SPI-3 series of (a) Kerala; (b) Telangana; (c) Orissa subdivisions and corresponding PAC plots. Upper panels shows the SPI-3 series and lower panels shows the corresponding PAC plots.

The sample PAC plots show that first four lags of SPI-3 series are significant in the data of all the three regions. Thus, the models take the functional form

$$SPI3(t) = f(SPI3(t-1), SPI3(t-2), SPI3(t-3), SPI3(t-4))$$

The four-input model is developed in the MEMD-SLR hybrid modeling framework by keeping 70% data for calibration and rest of the data for validation. The MEMD application resulted in 11 modes for the multivariate datasets of Kerala and Telangana while it resulted in 12 modes for the data from Orissa subdivision. SLR models are prepared for prediction of different rotary mode of SPI-3 of a generic time t, considering modes of predictor variables at the same scale as inputs. The resulting regression coefficient matrix for SPI-3 predictions in the three subdivisions Kerala, Telangana and Orissa are presented in Table 6.10 to Table 6.12.

From the regression coefficients of SPI-3 prediction for the three subdivisions (Tables 6.10–6.12), it is noticed that the components of lag-1 SPI-3 series have a positive influence on the respective modes of SPI-3, except for IMF1. For Kerala subdivision, the coefficients of lag-2 and lag-4 components are having a negative influence in the respective modes of SPI(t). Here except for IMF1, lag-3 component has a negative influence on the respective components of SPI(t). In the other two subdivisions, such uniform pattern is not observed except for the inference made for the coefficients of components of lag-1 SPI-3 series. Also, from Tables 6.10–6.12 it is noticed that at certain process scales, certain input variables are not influential, for example in IMF5 of SPI-3 of Telangana SPI-3(t-3) and SPI-3 (t-4) are not influential; IMF6 of Orissa, SPI-3(t-2) and SPI-3 (t-3) are not influential. It is to be noted that in

TABLE 6.10
Regression Coefficient Matrix for SPI-3 Prediction of Kerala Subdivision

Mode number of SPI3(t)	Mode number of			
	SPI3(t-1)	SPI3(t-2)	SPI3(t-3)	SPI3(t-4)
IMF1	−1.421	−1.312	−1.079	−0.506
IMF2	0.865	−1.549	0.607	−0.650
IMF3	2.021	−2.097	1.044	−0.365
IMF4	2.019	−1.525	0.453	−0.104
IMF5	2.033	−1.379	0.411	−0.125
IMF6	2.284	−1.654	0.351	*−0.024*
IMF7	1.616	−0.840	0.663	−0.402
IMF8	1.445	−0.566	0.350	−0.214
IMF9	2.215	−2.123	0.908	*−0.049*
IMF10	5.105	−6.542	3.716	−1.348
Residue	3.581	−3.981	1.800	−0.428

Note: The numbers in italics show that these coefficients are not significant at 5% significance level.

TABLE 6.11
Regression Coefficient Matrix for SPI-3 Prediction of Telangana Subdivision

Mode number of SPI3(*t*)	Mode number of			
	SPI3(*t-1*)	SPI3(*t-2*)	SPI3(*t-3*)	SPI3(*t-4*)
IMF1	−1.720	−1.699	−1.236	−0.495
IMF2	0.637	−1.429	0.422	−0.605
IMF3	2.001	−2.319	1.341	−0.506
IMF4	2.074	−1.525	0.320	*0.031*
IMF5	1.880	−0.939	*−0.001*	*0.004*
IMF6	1.742	−0.863	0.279	−0.191
IMF7	1.502	−0.272	−0.234	*−0.010*
IMF8	1.566	−0.880	0.350	*0.018*
IMF9	1.544	−0.769	0.390	−0.184
IMF10	0.229	2.826	−1.352	−0.771
Residue	0.162	2.937	−1.303	−0.860

Note: The numbers in italics show that these coefficients are not significant at 5% significance level.

TABLE 6.12
Regression Coefficient Matrix for SPI-3 Prediction of Orissa Subdivision

Mode number of SPI3(*t*)	Mode number of			
	SPI3(*t-1*)	SPI3(*t-2*)	SPI3(*t-3*)	SPI3(*t-4*)
IMF1	−1.871	−1.990	−1.469	−0.586
IMF2	0.908	−1.521	0.627	−0.586
IMF3	1.947	−1.896	0.816	−0.245
IMF4	2.107	−1.574	0.355	*0.017*
IMF5	1.972	−1.000	−0.209	0.193
IMF6	1.277	*−0.176*	*−0.001*	−0.087
IMF7	1.576	−0.737	−0.064	0.110
IMF8	1.561	−1.284	0.870	−0.175
IMF9	1.077	0.708	−1.879	1.083
IMF10	3.414	−2.512	4.891	−4.759
IMF11	1.692	−1.183	3.226	−2.737
Residue	0.429	0.701	−3.747	3.588

Note: The numbers in italics show that these coefficients are not significant at 5% significance level.

the present study only lagged SPI series are considered as inputs and such omissions will be more apparent if more input parameters including the teleconnected climate indices are used for prediction of SPI. Further in order to assess the efficacy of the method, MLR, M5 model tree, Genetic Programming (GP) and the MEMD-GP hybrid model are also developed to predict SPI-3 by considering the same input parameters. In the MEMD-GP hybrid model, instead of SLR, GP is used to model each of the rotatory modes of SPI-3 of the generic time t considering the rotatory modes of input variables of the same time scale as inputs. In GP modeling a function set comprising basic operators and conditional statement (*if-else*) are used. The results of performance evaluation of different methods are presented in Table 6.13. Also the scatter plots of SPI-3 predictions by different methods are presented in Figure 6.5–6.7, respectively, for Kerala, Telangana and Orissa subdivisions.

From Table 6.13 it is noticed that the correlation statistics are much higher (> 0.95) in all cases for predictions using both of the hybrid MEMD-SLR methods when compared to that by other four methods and also the error statistics are considerably low. In addition, it is noticed that the MEMD-GP hybrid model performs marginally better than MEMD-SLR hybrid model. For example, the R value and RMSE of predictions of SPI-3 of validation data of Kerala subdivision, by MEMD-GP model is 0.982 and 0.141 against 0.978 and 0.161, respectively, by the MEMD-SLR model. It is to be noted that only few regression coefficients are brought to zero during the MEMD-MLR model (Table 6.10), which infers all of the input parameters are influential at different time scales in the prediction of SPI-3, which is reflected in the improved performance of MEMD-GP model. The R^2 statistics of predictions of first three IMFs (which are high frequency IMFs) for the validation data of Kerala subdivision by the MEMD-GP method are 0.887, 0.947 and 0.984, respectively, while the values are 0.837, 0.897 and 0.946 by the MEMD-SLR method. It is to be noted that all the high frequency modes are more deterministic in nature, and the modeling

TABLE 6.13
Performance Evaluation of Different Models for SPI-3 predictions

Subdivision	PEC	MEMD-SLR		MEMD-GP		MT		MLR		GP	
		C	V	C	V	C	V	C	V	C	V
Kerala	R	0.978	0.978	0.983	0.982	0.737	0.640	0.656	0.650	0.679	0.639
	RMSE	0.211	0.209	0.183	0.188	0.680	0.751	0.755	0.743	0.734	0.753
	MAE	0.160	0.161	0.139	0.145	0.531	0.594	0.593	0.576	0.576	0.587
Telangana	R	0.983	0.984	0.985	0.985	0.751	0.632	0.677	0.661	0.700	0.662
	RMSE	0.183	0.170	0.172	0.172	0.652	0.743	0.724	0.716	0.702	0.717
	MAE	0.137	0.130	0.128	0.128	0.499	0.578	0.557	0.554	0.547	0.551
Orissa	R	0.979	0.976	0.982	0.977	0.740	0.651	0.645	0.698	0.677	0.664
	RMSE	0.200	0.217	0.186	0.213	0.658	0.763	0.744	0.725	0.718	0.754
	MAE	0.151	0.161	0.141	0.154	0.511	0.593	0.576	0.567	0.552	0.587

Note: PEC – Performance Evaluation Criteria; C – Calibration; V – Validation.

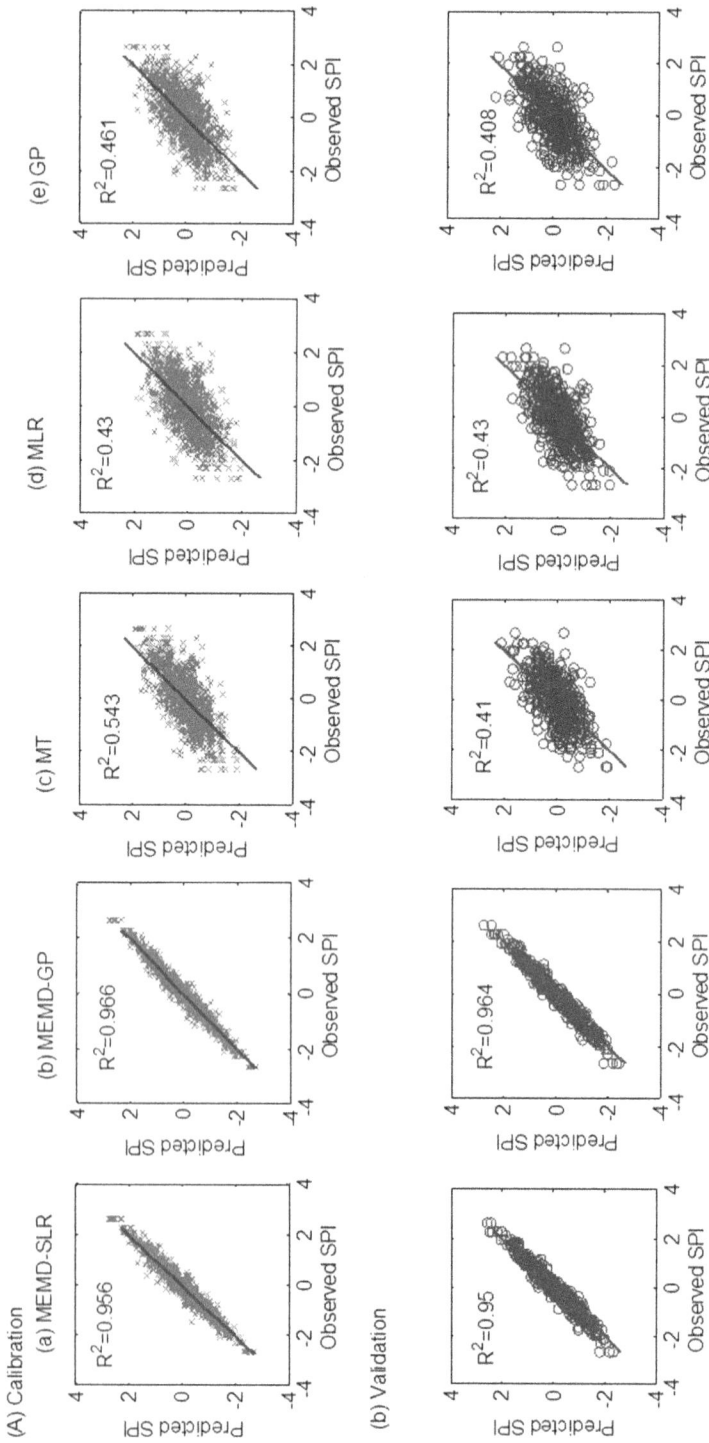

FIGURE 6.5 Scatter plots of SPI-3 predictions of Kerala subdivision by different models for calibration and validation datasets. Upper panels show the scatter plots for calibration dataset and lower panels shows the scatter plots for validation dataset.

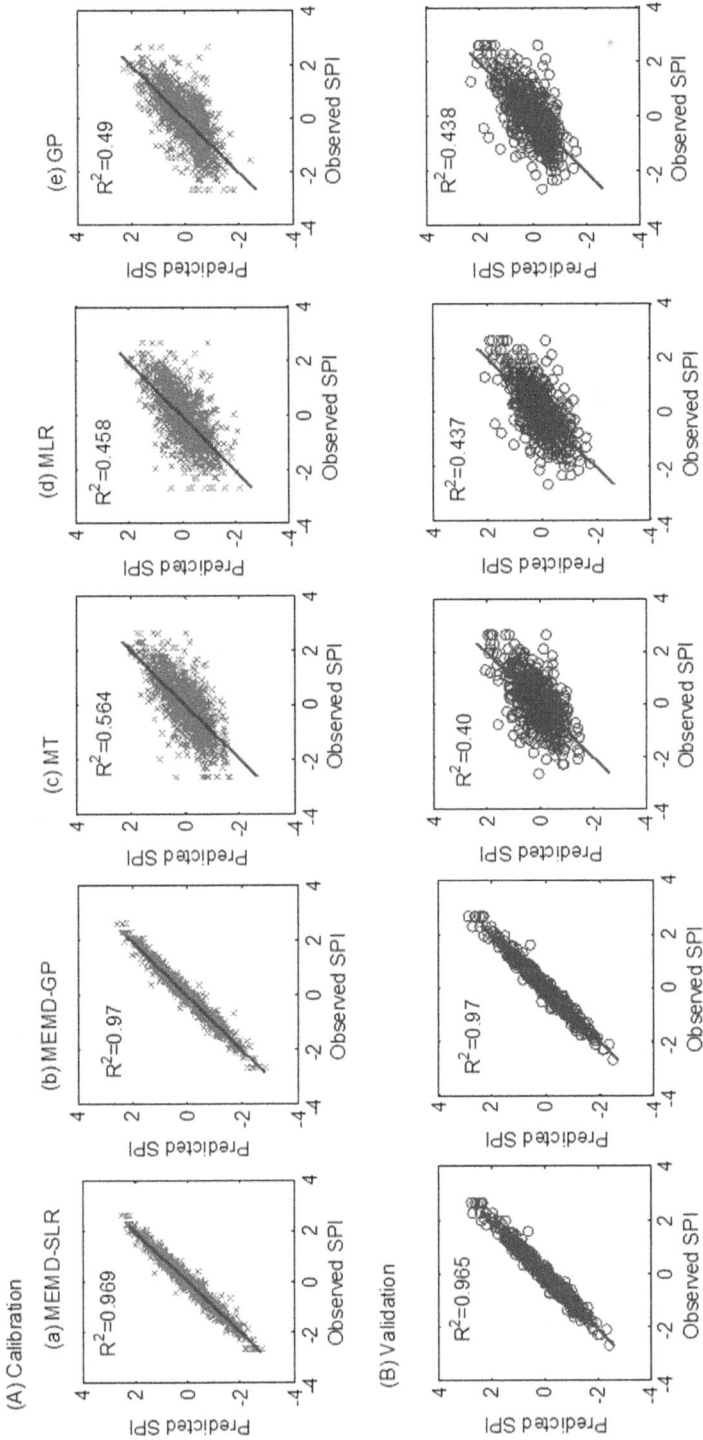

FIGURE 6.6 Scatter plots of SPI-3 predictions of Telangana subdivision by different models for calibration and validation datasets. Upper panels show the scatter plots for calibration dataset and lower panels shows the scatter plots for validation dataset.

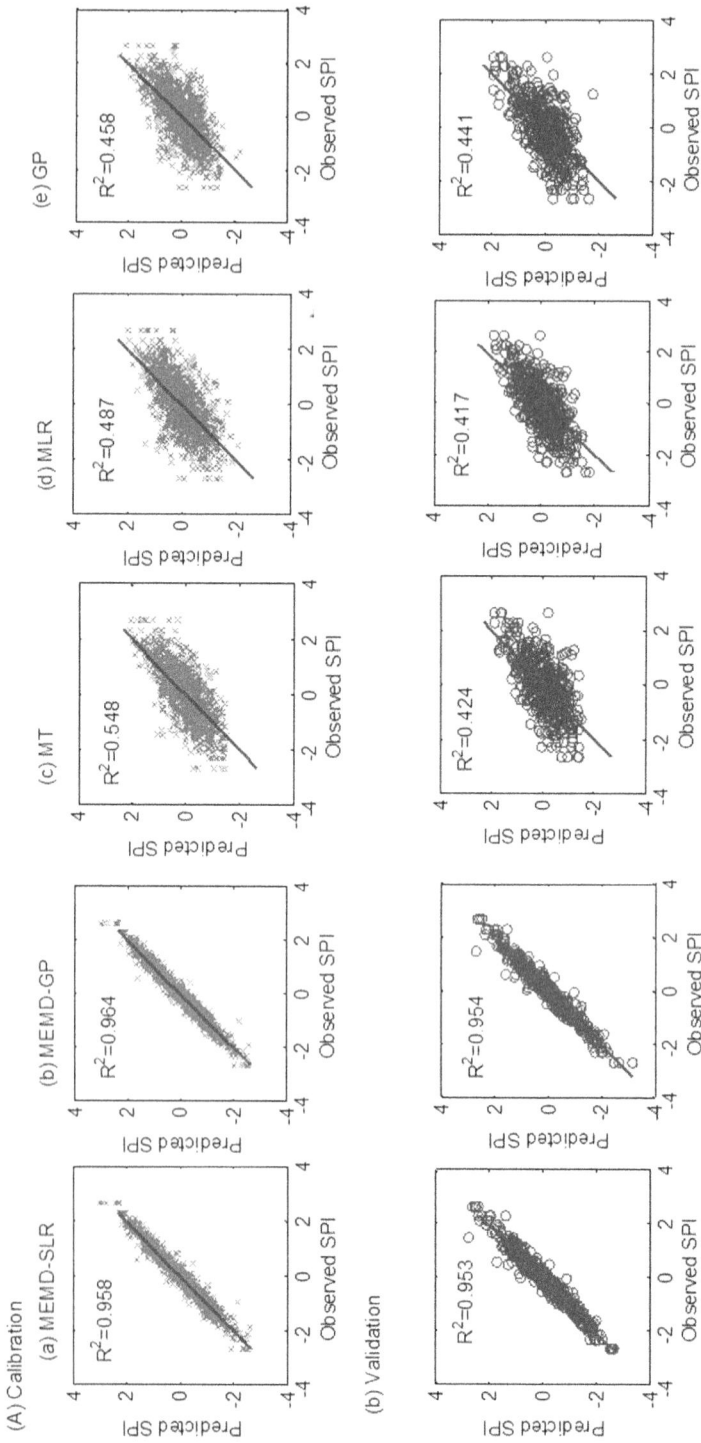

FIGURE 6.7 Scatter plots of SPI-3 predictions of Orissa subdivision by different models for calibration and validation datasets. Upper panels shows the scatter plots for calibration dataset and lower panels shows the scatter plots for validation dataset.

of these components resulted in R^2 statistics of more than 0.98 consistently, by both GP and SLR methods. Hence the MEMD-GP method is better in modeling the high frequency modes than MEMD-SLR method, which may be the reason for marginal improvement in performance of overall prediction for SPI-3 using the former method when compared to the latter one. Thus, it is established MEMD-based hybrid models are effective in modeling short term droughts when compared with the linear and nonlinear models.

6.4 RESERVOIR INFLOW PREDICTION USING THE MEMD-SLR MODEL

The climate teleconnection studies on reservoir inflows proved that such associations are not of unique character at different time scales, but they differ in both the strength and nature of association at different process scales (Adarsh and Janga Reddy 2019). It is believed that capturing such information may improve the accuracy of inflow predictions rather than proceeding with conventional linear/nonlinear time series methods. Hence, MEMD-SLR approach is implemented to predict monthly inflow to Hirakud reservoir. In this exercise, first the multivariate dataset comprising the monthly inflow time series of Hirakud reservoir of 1958–1993 along with the two potential climate variables ENSO and EQUINOO of the same period is decomposed using MEMD. The minimum and maximum threshold parameters are set as 0.075 and 0.75, respectively, while the fraction controlling the sifting iterations is fixed as 0.075, following the recommendation from the past studies (Hu and Si 2013). The decomposition resulted in six modes and residue and the lagged values of inflow along with the lagged values of ENSO and EQUINOO indices are used as inputs to predict monthly reservoir inflows. To identify an appropriate predictor set, Partial Auto Correlation Function (PACF) of monthly inflows is computed and presented in Figure 6.8.

From the PACF plot of inflows it is noticed that three input lagged values influences the inflow of any generic time t. In some of the earlier studies upto two lagged values of ENSO index and three lagged values of EQUINOO index parameters are

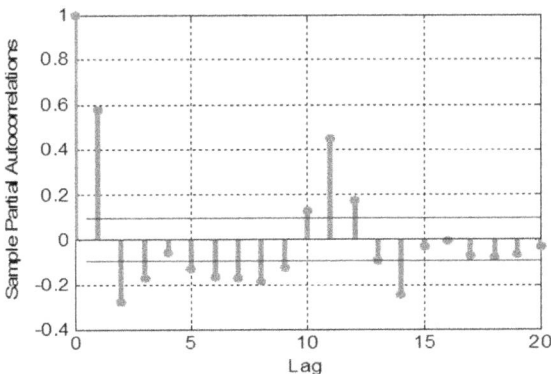

FIGURE 6.8 PACF plot of monthly inflows to Hirakud reservoir.

considered in fixing model inputs of monsoon streamflow predictions of Mahanadi river (Maity and Nagesh Kumar 2008, 2009). Hence, three lagged values of inflow (Q_{t-k}) $(k = 1,2,3)$, two lagged values for ENSO index (EN_{t-k}) $(k = 2)$ and three lagged values of EQUINOO index (EQ_{t-k}), $(k = 1,2,3)$ where k is the index of the lag chosen for the analysis for each predictor variable to demonstrate the utility of the proposed methodology for inflow prediction. The above 8 predictors and inflow of time t constituted the multivariate dataset for the present problem. Hence the functional form of the model can be represented as

$$Q_t = f(Q_{t-1}, Q_{t-2}, Q_{t-3}, EN_{t-1}, EN_{t-2}, EQ_{t-1}, EQ_{t-2}, EQ_{t-3})$$

In the modeling using MEMD-SLR approach, the model of individual rotational mode takes the general functional form

$$RM_{iQ_t} = f \left(\begin{array}{c} RM_{iQ_{t-1}}, RM_{iQ_{t-2}}, RM_{iQ_{t-3}}, RM_{iEN_{t-1}}, RM_{iEN_{t-2}}, \\ RM_{iEQ_{t-1}}, RM_{iEQ_{t-2}}, RM_{iEQ_{t-3}} \end{array} \right) \text{ and}$$

$$Q_t = \sum_{i=1}^{N} RM_i \text{ where } i \text{ is the index of the orthogonal mode;}$$

and N is the number of rotational modes.

However, it is to be noted that experimentation with multiple combination of inputs can be performed for the inflow prediction, as the methodology proposed is a general one.

The multivariate dataset is decomposed by MEMD method and the decomposition resulted in 8 IMFs and residue for each variable. The SLR models are fitted for individual IMFs and the residue of $Q(t)$. In the implementation of SLR, such components which have the p-value less than 0.05 are excluded. To illustrate this point, the initial coefficient matrix along with the final fitting statistics (after bringing the insignificant coefficients to zero) is shown in Table 6.14. It is to be noted that the numbers in italics show that p-value corresponding to these coefficients exceeds 0.05 and hence not acceptable for regression modeling.

From Table 6.14, it is noticed that at the time scale of IMF1, both of the lagged values of ENSO are not influential (hence ENSO plays no role in deciding the streamflow magnitude at this time scale). Similarly, EQUINOO and inflow of lag 3 (Q_{t-3}) are not influential at time scale of IMF3. Therefore, such coefficients are to be brought zero to prepare the final SLR models. The fitting statistics shown in Table 6.6 corresponds to the final SLR models.

For example, the final SLR model for prediction of IMF1 can be represented as

$$IMF1_{Q_t} = -1.015 IMF1_{Q_{t-1}} - 0.89 IMF1_{Q_{t-2}} - 0.464 IMF1_{Q_{t-3}} - 560.25 IMF1_{EQ_{t-1}}$$
$$- 1157.42 IMF1_{EQ_{t-2}} - 750.04 IMF1_{EQ_{t-3}} + 71.99$$

with an R^2 of 0.676, RMSE of 1123 and F-statistic of 100.

TABLE 6.14

Regression Coefficient Matrix of Models Fitted to Predict the Rotational Modes of Q_t Along with the Fitting Statistics

Mode Number of Q_t	Input Variable (IMF of)								Intercept	R^2	RMSE	F-Statistics
	Q_{t-1}	Q_{t-2}	Q_{t-3}	EN_{t-1}	EN_{t-2}	EQ_{t-1}	EQ_{t-2}	EQ_{t-3}				
IMF1	−1.01	−0.89	−0.46	−1274.2	*1118.3*	−560.25	−1157.4	−750.04	71.99	0.676	1123.35	100.04
IMF2	0.36	−0.52	−0.21	*−128.34*	−1461.13	553.36	−637.25	918.71	−6.36	0.873	887.797	280.96
IMF3	1.61	−0.98	*0.001*	−1743.76	1605.43	*125.16*	*88.18*	*10.86*	*10.61*	0.977	622.628	3107.22
IMF4	1.69	−0.58	−0.26	−1171.82	1104.33	−643.53	148.23	663.54	−8.28	0.994	92.811	6815.07
IMF5	*0.18*	*1.706*	−0.82	−458.02	370.05	1660.84	−4552.8	3446.23	1.64	0.990	76.702	4218.43
IMF6	2.57	−2.74	1.00	1577.87	−1299.97	−635.93	1685.27	−1039.5	−1.03	0.997	21.668	10960
IMF7	1.47	−1.46	0.883	−1610.85	1164.445	−7140.4	4102.37	3359.28	0.866	0.999	1.906	2.29×10^{10}
IMF8	1.05	0.21	−0.25	1854.882	−1973.68	105.96	1064.18	−1238.1	−0.013	1.000	0.001	2.89×10^{10}
Residue	0.91	0.56	−0.46	0.000	0.000	169.739	0.000	0.000	−34.24	1.000	0.001	2.9×10^{10}

Note: The numbers in italics indicate the p-value is not significant and the coefficients are to be brought to zero in the models of different subseries.

Similarly, the equations for other modes can be obtained. It is noticed that the coefficient of determination (R^2) value increases with increase in mode number and the lowest R^2 values are noticed for the first two IMFs. Hence linear fitting works better for higher order modes, which is quite obvious, as the lower order IMF (say IMF1) is expected to contain noise. The *F*-statistics and RMSE also confirm the effectiveness of fitted LR models. The predicted modes are added to get the inflow at time t (Q_t). The different models are used to predict the modes of Q_t at respective time scales for the validation data. The final summation of predicted modes gives the Q_t for validation data. To evaluate the efficacy of the proposed approach, the models using M5 Model Tree (MT) algorithm, MLR, GP, ANN are developed by considering the same predictor set (lagged inputs of streamflow, ENSO, EQUINOO) to predict Q_t. To develop the ANN model, an architecture of 8-6-1, the number of neurons in the hidden layer of 6, learning rate of 0.5, momentum coefficient of 0.6, hyperbolic tangent sigmoid transfer function, resilient back-propagation learning algorithm etc. used and these parameters are chosen by a trial and error approach. In the application of GP, a population size of 100, initial mutation frequency of 0.9 and crossover frequency of 0.5, a function set comprising *if-else* rules along with basic arithmetic operators (*, +, −) are utilized. The observed and predicted streamflow for calibration and validation datasets by the different methods are presented in Figure 6.9, and the scatter plot of predictions for calibration and validation datasets are presented in Figure 6.10. Different evaluation measures such as R, NSE, RMSE, MAE and Scatter Index (SI) are considered to examine the performance of different models.

The performance measures for results obtained by different models (both for calibration and validation datasets) are summarised in Table 6.15.

From Table 6.15, it is clear that the MEMD-SLR method performs better in inflow prediction than the MLR and other data-driven methods both in terms of error statistics and the correlation measures. The results in Table 6.7 show that R^2 and NSE values of predictions during validation period are the highest for MEMD-SLR model. The RMSE statistics for predictions during validation period are smaller, which is 48% better than the best performing MT approach, which is obviously a significant improvement. Further to evaluate its capabilities of peak flow prediction, the flows exceeding (mean+2*SD) are considered in calibration data and 10 points are identified in validation data. Moreover, the capabilities of models in predicting low as well as medium flows also examined. Since the standard deviation of the calibration and validation data (5170 and 4180 m³/sec, respectively) is higher than the mean values (2180 and 2650 m³/sec, respectively) all the flow lower than the mean is categorised as low flow. The calibration data constitute 220 low flow points and 53 medium flow points, while validation data constituted 91 low flow points and 25 medium flow points. The observed and predicted flows by three methods for different flow states (high, low and medium) for calibration and validation periods are presented in Figure 6.11. The performance of predictions of different flow states (HF, LF and MF) are statistically evaluated and presented in Table 6.16.

From the results presented in Table 6.16 it can be noticed that MEMD-SLR hybrid model has resulted in much less error statistics for inflow prediction as compared with GP, ANN, MT and MLR models for different flow states. The statistics in Table 6.16 further show that MEMD-SLR model is also capable in predicting the peak flows quite well when compared with ANN, GP, MT and MLR models.

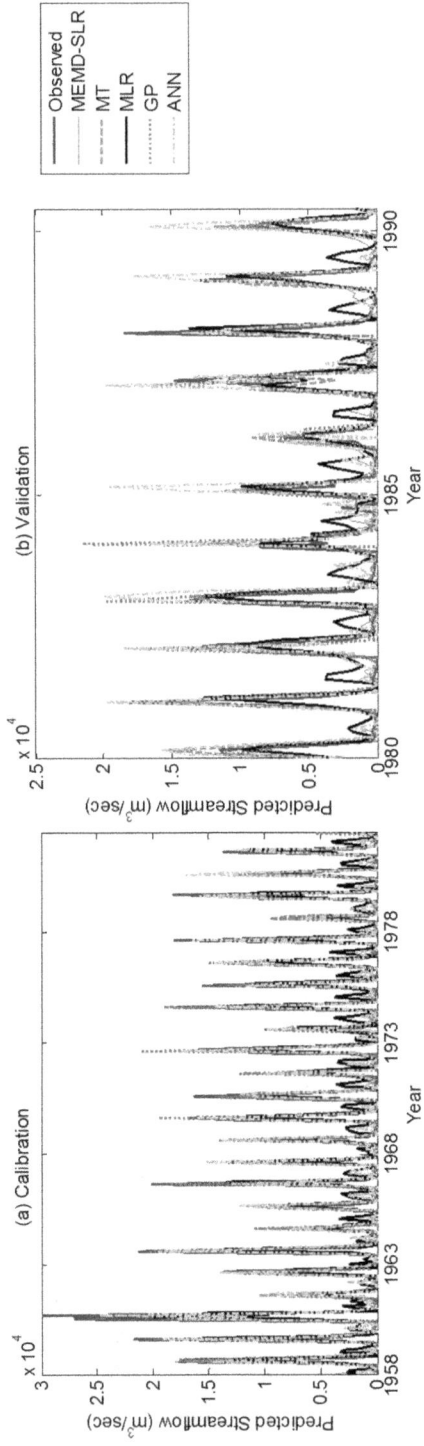

FIGURE 6.9 Time series plots of observed and predicted inflow for calibration and validation datasets (From Adarsh 2017).

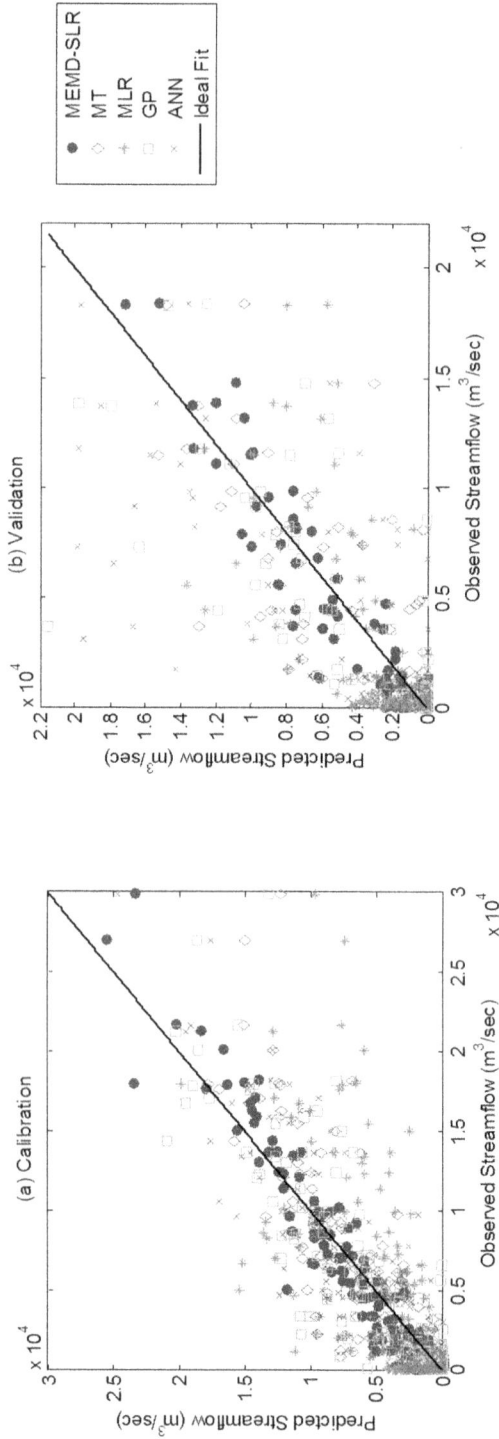

FIGURE 6.10 Scatter plots of inflow predictions by different methods for calibration and validation datasets. (From Adarsh 2017).

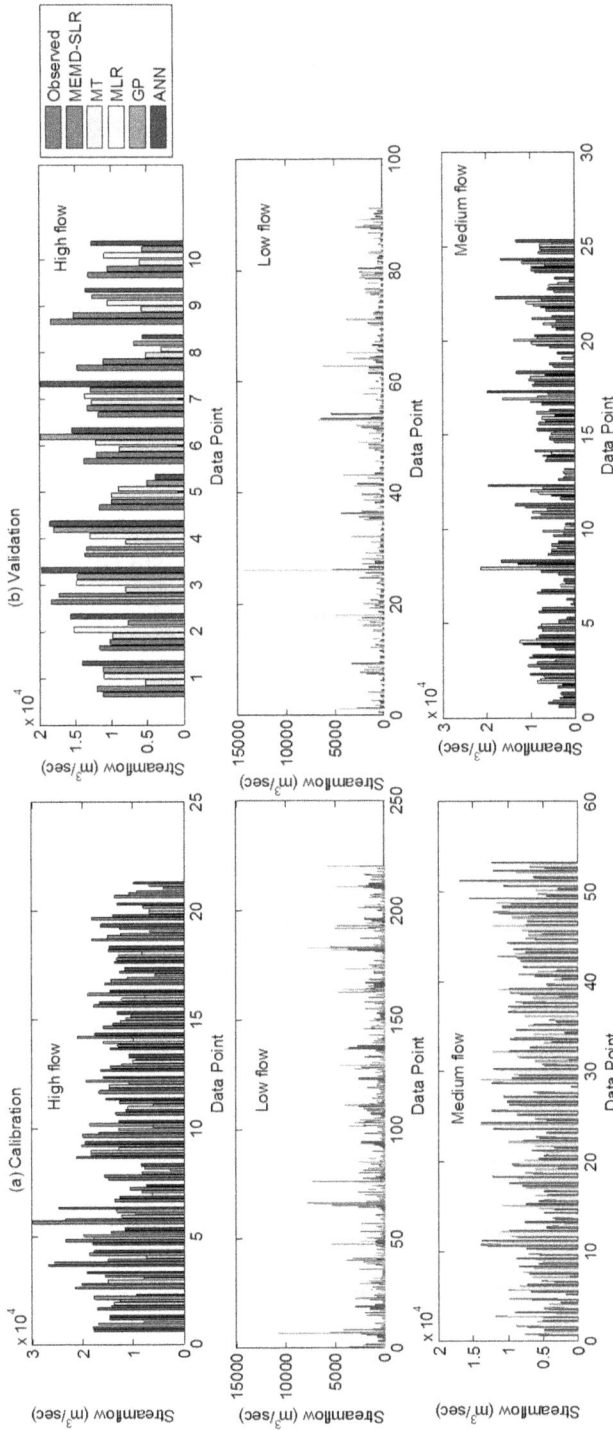

FIGURE 6.11 Comparison of observed and predicted flows by different methods during calibration and validation periods. (From Adarsh 2017).

TABLE 6.15
Performance Evaluation of Different Models for Reservoir Inflow Prediction

PEC	Calibration data					Validation data				
	MEMD-SLR	MT	MLR	GP	ANN	MEMD-SLR	MT	MLR	GP	ANN
R^2	0.939	0.812	0.458	0.901	0.910	0.905	0.667	0.444	0.778	0.756
NSE	0.935	0.803	0.458	0.811	0.827	0.901	0.647	0.429	0.493	0.541
RMSE	1314.16	2294.8	3803.4	2241	2150	1307.88	2472.2	3144.4	2964.	3695
MAE	820.11	1131.12	2509.96	1071.2	1334.9	938.81	1363.9	2156.17	1486.8	2119.5
SI	0.468	0.817	1.354	0.797	0.765	0.493	0.932	1.186	1.117	1.027

Note: PEC refers to performance evaluation criteria.

TABLE 6.16
Performance Evaluation of Different Models for Different Flow States

Flow state	PEC	Calibration data					Validation data				
		MEMD-SLR	MT	MLR	GP	ANN	MEMD-SLR	MT	MLR	GP	ANN
HF	R	0.852	0.335	0.154	0.314	0.708	0.780	0.041	0.383	0.322	0.171
	RMSE	2768.78	6694.47	10507.4	6117.54	4813.91	2097.30	4953.65	7128.03	5241.36	5358.06
LF	R	0.669	0.648	0.384	0.648	0.447	0.435	0.624	0.323	0.406	0.422
	RMSE	921.381	1182.78	2000.70	1128.04	1186.93	992.657	1262.44	1933.91	1159.37	1881.00
MF	R	0.806	0.679	0.249	0.599	0.581	0.669	0.190	-0.028	-0.020	0.148
	RMSE	1737.43	2374.35	4279.45	2750.50	3258.93	1809.69	3896.95	3986.66	5326.64	6667.64

Note: PEC refers to performance evaluation criteria; HF – High Flow; LF – Low Flow; MF –Medium Flow.

6.5   PREDICTION OF DAILY SUSPENDED SEDIMENT LOAD CONCENTRATION

Modelling suspended sediment load from rivers is a complex problem in hydrology, which plays a significant role in river basin management, assessment of risk, operation and monitoring of hydraulic structures like dams and canals. Hydrological time series often possess nonstationarity characteristics, where the statistical properties like mean and variance may vary in the time domain. These changes may be due to several factors such as climate change, human interventions like construction and operation of storage or diversion structures, changes in the method of measurement etc. may induce nonstationarity in hydrological time series. In general, the time series approach for hydrologic prediction involves an inherent assumption of stationarity, i.e., the pattern of historical records will be reflected in future also. Under a nonstationary environment, such modeling approach may fail, as the time series models developed by traditional approaches or data-driven paradigms works well only for the range of datasets chosen and their generalization capability will be poor if a data of drastically different statistical properties are supplied for predictions. Therefore, the normal way of data-partitioning into training and validation may lead to erroneous results. The analysis of daily streamflow and suspended sediment load data from Kallada river using CWT detected a change about the year 2000 (Chapter 3). Furthermore, local investigations showed that measurement method changed in 2009–2010 period and a reduction in precipitation events is noticed in some of the earlier studies. The daily time series data of streamflow and suspended sediment flux are provided in Figure 6.12. Moreover, the basic examination of statistical properties of both the time series is performed after partitioning the data of 2002–2009 for calibration and 2009–2012 for validation. The results are presented in Table 6.17.

From Table 6.17 it is clear that from the statistical properties (mean and standard deviation) of both the time series; the properties for validation data are considerably larger than that for respective calibration dataset. The mean of validation data of SSF time series is three times and SD is six times larger than that of calibration data.

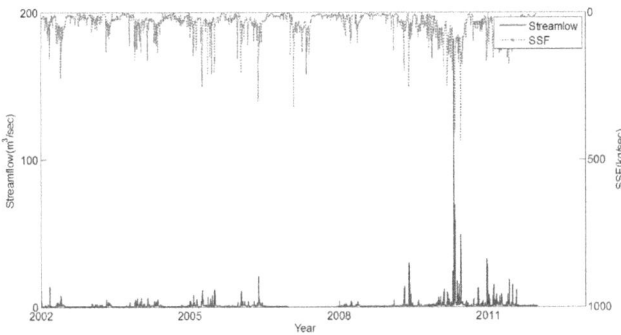

FIGURE 6.12   Daily streamflow and suspended sediment concentration data from Pattazhy station (2002–2012).

TABLE 6.17
Statistical Properties of Daily Streamflow and Suspended Sediment Flux from Kallada River (2002–2012 Period) for Calibration and Validation Data

Statistical property	Calibration		Validation	
	Streamflow (m³/sec)	Suspended sediment flux (kg/sec)	Streamflow (m³/sec)	Suspended sediment flux (kg/sec)
Maximum	322.939	20.201	502.619	181.445
Minimum	0.449	0.000	2.882	0.007
Mean	28.818	0.513	52.780	1.701
Standard deviation	29.428	1.113	50.362	7.260

From the data range (min-max) it is clear that during the validation period, the values beyond the range used for calibration need to be predicted (based on the calibrated model), these may give erroneous results on proceeding with conventional time series modeling methods. In the next section, the potential of the MEMD-SLR method is evaluated in this perspective.

For modelling SSL, time series analysis is used as a simple and popular approach based on the inherent assumption that the lagged values of suspended sediment load and streamflow fairly represents the factors deciding the magnitude of suspended sediment load at the present time step. Like many other predictive modelling problems in hydrology, the application of numerous data driven techniques and their hybrid variants have been proposed in the past for the prediction of suspended sediment load (Kisi et al. 2008; Rajaee et al. 2009, 2011; Senthil Kumar et al. 2011; Lafdani et al. 2013; Haji et al. 2014; Nourani and Andalib 2015; Zounemat-Kermani et al. 2016). In this section the proposed hybrid approach involving MEMD and SLR model is invoked for modeling daily suspended sediment load of Kallada river in Kerala, India by considering the data for 2002–2009 for calibration and 2009–2012 for validation. In the modeling process, first the PAC plot is prepared for suspended sediment flux (SSF) time series data for identifying appropriate inputs. As the streamflow is the major driver for suspended sediment in natural rivers, the correlation of the sediment load series of current time step with the lagged streamflow values were estimated, and corresponding plots are presented in Figure 6.13.

From Figure 6.13, it is noticed that first three lags of SSF show reasonable correlation; and also three lags of streamflows are found to be most correlated with SSF. Therefore, to model SSF and to examine the effect of streamflow on the variability of SSF, five different model combinations of input datasets are considered. The input-output combination for the different models (Models 1 to 5, respectively) are:

$$Model 1: \quad SSF_t = f(Q_{t-1}, SSF_{t-1})$$

$$Model 2: \quad SSF_t = f(SSF_{t-1}, SSF_{t-2})$$

$$Model 3: \quad SSF_t = f(Q_{t-1}, SSF_{t-1}, SSF_{t-2})$$

$$Model 4: \quad SSF_t = f(Q_{t-1}, Q_{t-2}, SSF_{t-1}, SSF_{t-2})$$

$$Model 5: \quad SSF_t = f(Q_{t-1}, Q_{t-2}, Q_{t-3}, SSF_{t-1}, SSF_{t-2}, SSF_{t-3})$$

where SSF is the suspended sediment flux (in kg/sec) and Q is the streamflow (in m^3/sec).

The MEMD method is invoked upon the multivariate dataset prepared for each case. Here a control parameter set of 0.075, 0.75 and 0.075; and direction vector of 64 are chosen for MEMD implementation after following the suggestions given in the past studies (Hu and Si 2013). The decomposition resulted in 11 modes including residue for models 1 and 4; 12 modes for models 2 and 5 while for model 2, the decomposition resulted in 14 modes. It is to be noted that the number of modes

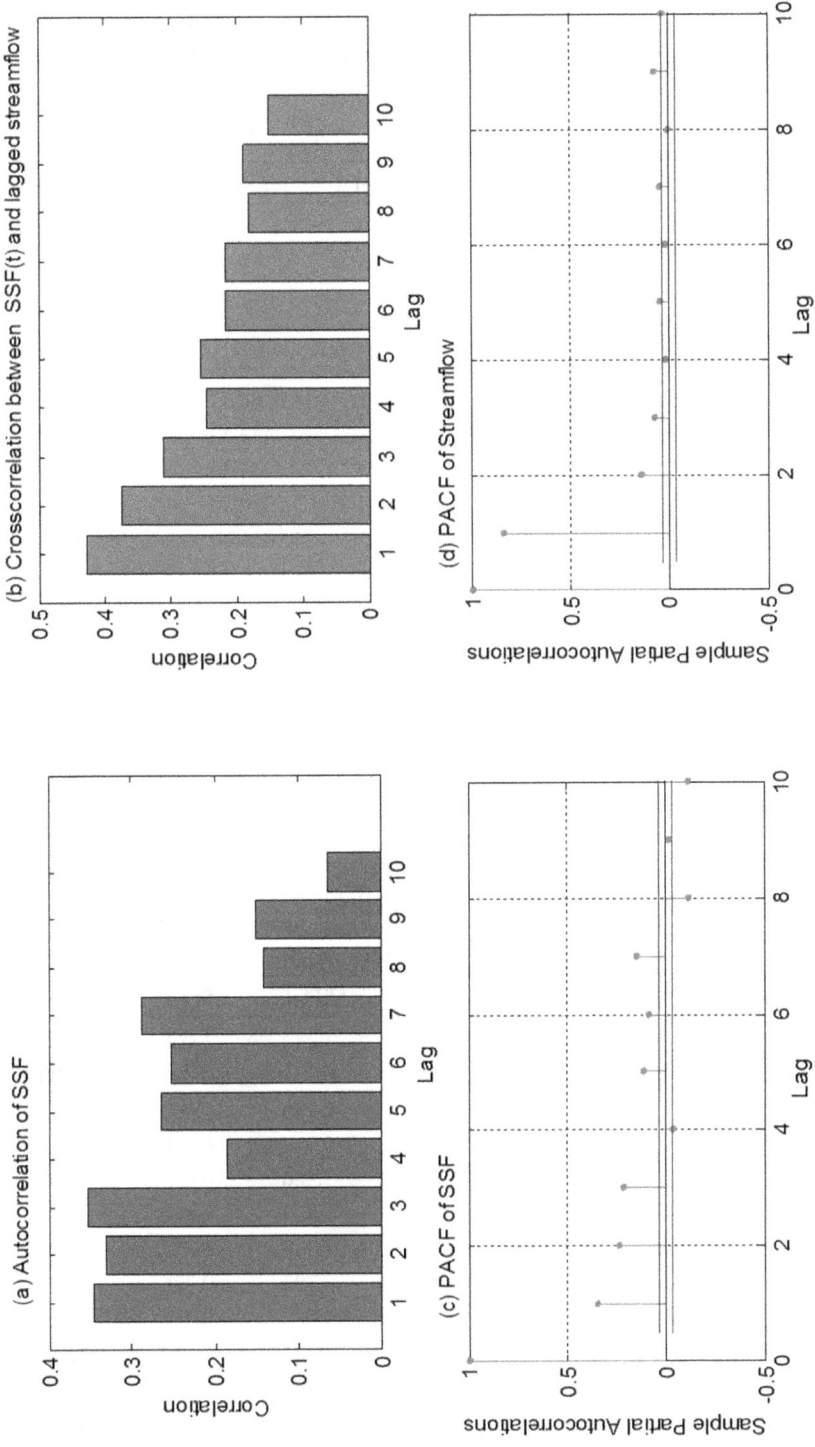

FIGURE 6.13 Autocorrelation and PAC of streamflow and suspended sediment flux time series from Pattazhy station.

TABLE 6.18
Mean Period (in days) of Multivariate Datasets of Different Models for SSF Prediction for Pattazhy Station

Mode Number	Mean Period (days)				
	Model 1	Model 2	Model 3	Model 4	Model 5
1	3.42	3.41	3.45	3.28	3.01
2	6.36	5.56	6.43	6.01	5.22
3	12.81	9.58	12.74	11.14	9.58
4	24.45	17.74	25.00	21.13	17.31
5	50.60	31.22	47.79	40.66	33.07
6	96.80	54.91	96.49	86.37	64.60
7	185.95	86.04	224.28	172.78	116.67
8	404.67	151.75	547.38	333.30	212.54
9	849.80	331.09	864.98	910.50	426.35
10	1618.67	520.29	1214.00	1821.00	910.50
11	LT	1011.67	3186.75	LT	1821.00
12		1821.00	LT		LT
13		3642.00			
14		LT			

Note: LT refers to long term.

TABLE 6.19
Results of MEMD-SLR Modeling Shows Regression Coefficient Matrix for Prediction of Daily SSF of Kallada River (Model 5)

Mode of SSFt	Mode of					
	Qt-1	Qt-2	Qt-3	SSFt-1	SSFt-2	SSFt-3
IMF1	0.003	0.000	0.000	−1.282	−1.066	−0.553
IMF2	0.000	0.000	0.000	0.435	−0.705	0.000
IMF3	0.011	−0.019	0.014	0.796	0.000	−0.567
IMF4	0.006	0.000	−0.005	0.946	0.478	−0.738
IMF5	−0.036	0.075	−0.037	1.586	−0.655	0.000
IMF6	−0.049	0.090	−0.040	1.849	−0.860	0.000
IMF7	−0.066	0.141	−0.078	1.949	−1.325	0.370
IMF8	0.031	−0.052	0.019	−0.793	0.726	1.081
IMF9	0.021	0.000	−0.017	0.000	−0.333	1.227
IMF10	0.172	−0.253	0.078	−1.784	2.214	0.547
IMF11	0.049	−0.164	0.112	−0.373	1.324	0.085
Residue	−0.025	−0.047	0.066	0.766	0.330	0.000

evolved may also depend on the characteristics of the dataset used (Huang et al. 2009a). The mean periods of the multivariate datasets of different models computed by zero crossing method (Huang et al. 2009b) are provided in Table 6.18.

In all cases, separate regression models are fitted to predict each rotational mode of SSL using the SLR method considering the rotational modes of predictor variable at the corresponding time scale as inputs. The regression coefficient matrix is presented in Table 6.19.

From Table 6.19, it can be noted that the coefficients for streamflow are close to zero in many of the models, which shows lesser influence of streamflow in modeling SSF. In other words, lagged SSF only are sufficient to model SSF of the generic time step t.

To examine the relative efficacy of MEMD-SLR model, the MT and MLR methods are also used for SSF predictions for different models (Models 1 to 5). The M5 model tree method is found to be superior to some of the other data driven techniques in suspended sediment modeling (Bhattacharya and Solomatine 2006; Senthil Kumar et al. 2011) and moreover the method is hardly biased by the algorithm specific control parameter settings (Singh et al. 2010). The performance evaluation of different methods in modeling daily SSF for different models is presented in Table 6.20.

TABLE 6.20
Performance Evaluation of Different Models for SSF Predictions by Different Methods

Model No	PEC	MEMD-SLR		MT		MLR	
		C	V	C	V	C	V
Model1	R	0.782	0.729	0.700	0.323	0.755	0.388
	NSE	0.555	0.532	0.490	−0.001	0.568	0.135
	RMSE	0.741	4.995	0.793	7.301	0.730	6.786
Model2	R	0.921	0.832	0.704	0.348	0.750	0.390
	NSE	0.844	0.687	0.495	0.021	0.562	0.139
	RMSE	0.438	4.082	0.789	7.219	0.735	6.771
Model3	R	0.857	0.829	0.705	0.355	0.755	0.394
	NSE	0.728	0.685	0.497	0.054	0.568	0.139
	RMSE	0.579	4.095	0.788	7.100	0.730	6.771
Model4	R	0.926	0.853	0.705	0.360	0.766	0.395
	NSE	0.856	0.725	0.497	0.063	0.585	0.139
	RMSE	0.421	3.829	0.788	7.065	0.716	6.770
Model5	R	0.958	0.882	0.706	0.374	0.770	0.395
	NSE	0.918	0.778	0.498	0.082	0.590	0.139
	RMSE	0.319	3.442	0.787	6.991	0.711	6.770

Note: PEC – performance evaluation criteria; C – Calibration; V – Validation.

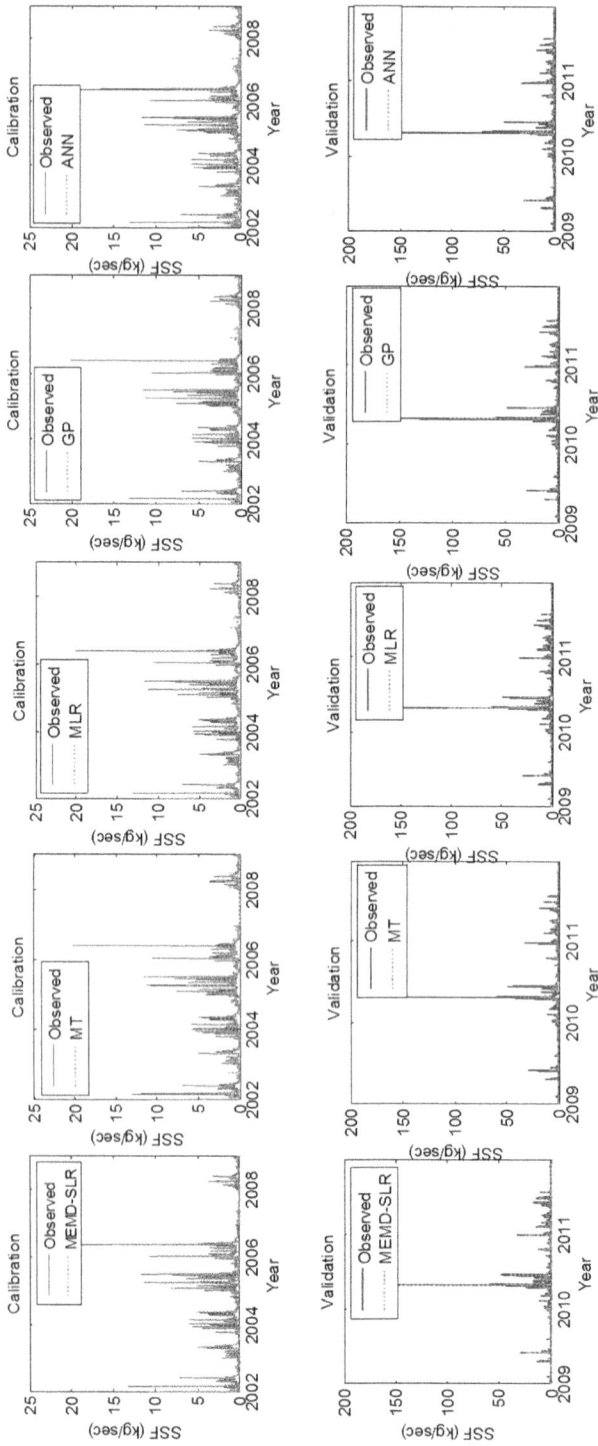

FIGURE 6.14 Time series plots of daily SSF predictions of Kallada river by different models (MEMD-SLR, MT, MLR, GP, ANN). The upper panels show the plots for calibration dataset and lower panels shows the plots for validation dataset.

An examination of the results based on multiple statistical performance evaluation criteria shows that the MEMD-SLR coupled approach is found to be performing better than MLR and M5 model tree (MT) in all of the five cases (Table 6.20). Among the different models considered in this study, the one that accounts three lagged values of SSF and streamflow displayed the best performance.

From Table 6.20, it can also be noted that the correlation statistics of validation data is much less for all the five models while using M5 model tree and MLR. This was quite expected on recollecting the fact that the statistical properties (mean and standard deviation) of validation data (2009–2012 period) are considerably larger than that for calibration dataset (from Table 6.10). So, in order to examine the predictability further, the nonlinear data driven methods ANN and GP are also applied to predict SSF considering the best performing input combination (of Model 5). To develop the ANN Model, an architecture of 6-10-1 is adopted, which has 10 neurons in the hidden layer, the learning rate of 0.5, momentum coefficient of 0.7, hyperbolic tangent sigmoid transfer function, resilient back propagation learning algorithm, etc. and these parameters are selected by a trial-and-error approach. In the application of GP, a population size of 100, initial mutation frequency of 0.9 and crossover frequency of 0.5, a function set comprising *if-else* rules along with basic arithmetic operators (*, +, −) are used. In this case, the correlation coefficients for validation data are 0.475 and 0.544; NSE values are 0.219 and 0.282, whereas the RMSE values are 6.42 and 6.18, respectively, for GP and ANN models. On examining the results of predictions by different methods (Table 6.13) it is noticed that the correlation coefficient and NSE of MEMD-SLR model are much better (0.882 and 0.778) than that for other data-driven models for the validation data. Also, the RMSE statistics of MEMD-SLR model is found to be ~44% better than that of best performing ANN model. Hence it is evident that the MEMD-SLR models display better predictive capabilities than that of data driven methods like GP, ANN, MT and the MLR method.

The time series plots and scatter plots of predictions by different methods are given in Figure 6.14 and Figure 6.15 respectively. For better visualization of results, the predictions of some of the selected data segments (Data points 350–400, 600–650

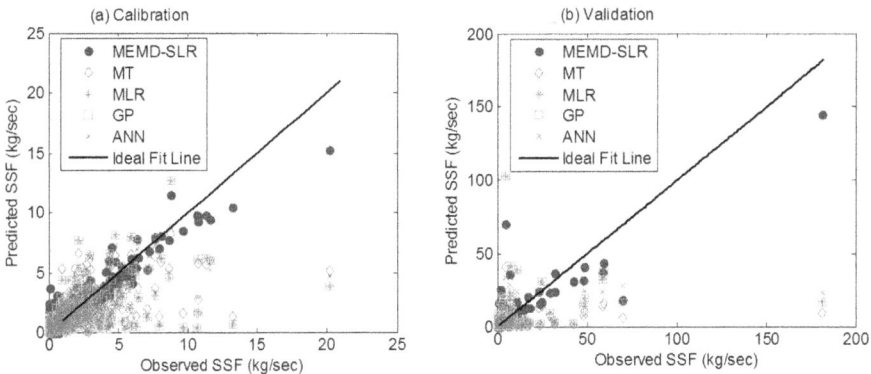

FIGURE 6.15 Scatter plots of daily SSF predictions of Kallada river during (a) calibration period (2002–2009) and (b) validation period (2009–2012).

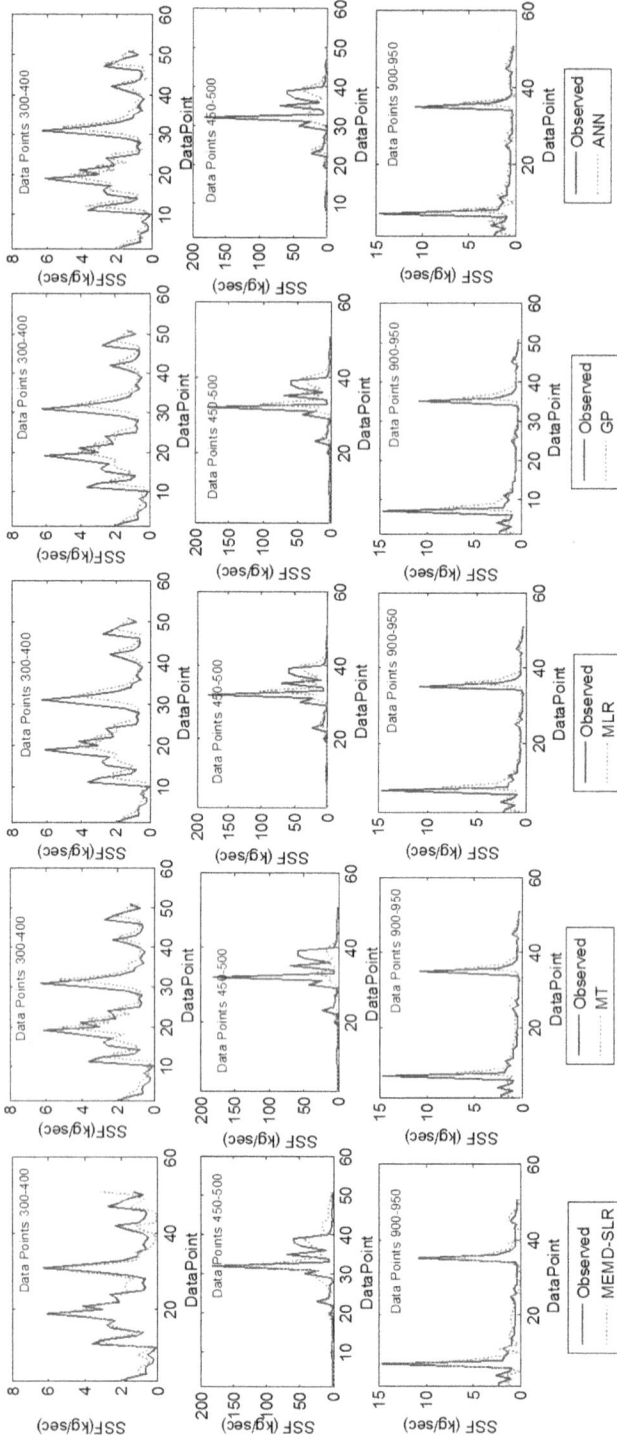

FIGURE 6.16 Predictions of SSF by different models for different segments (300–400, 450–500, 900–950) from validation dataset.

and 900–950 taken arbitrarily, which comprise higher values as well as lower values (exceeding 50 kg/sec, comprise 10–15 kg/sec; less than 10 kg/sec) are plotted and presented as Figure 6.16.

A careful perusal of y-axis of the time series plots (upper panels and lower panels of Figure 6.14) show that data range of validation data is beyond that for calibration data. Further, from the scatter plots (Figure 6.15), it is clear that the MEMD-SLR predictions are closer to the ideal fit line than that by other methods.

A careful examination of Figure 6.16 also clearly shows that MEMD-SLR model captures the variability in a better way than the other methods for datasets in different ranges (values exceeding 50 kg/sec, comprise 10–15 kg/sec; less than 10 kg/sec from y-axis of different panels of Figure 6.16). Hence, it can be concluded that the MEMD-SLR method displayed excellent generalization capability than the other data driven methods when applied for prediction of daily suspended sediment load of Kallada river.

6.6   CLOSURE

This chapter first presented different case study applications of MEMD-based hybrid models for hydrological simulations. The MEMD based hybrid models are developed by using MEMD as a preprocessing decomposition tool, following by integrating with linear or nonlinear regression methods. The decomposition enables us to get information from multiple process scales and MEMD offers the flexibility in accounting different input variables for hydrological simulations. The use of SLR helpful in retaining the most relevant inputs in different time scales, by automatically ruling out the less significant predictors. The application of MEMD-SLR hybrid model used a strategic approach for prediction of rainfall of typical month considering the appropriate lagged year rainfall values, along with the appropriate lagged values of two large scale climatic oscillations (ENSO and EQUINOO). Here the past experience and statistical correlations were considered in fixing the inputs. The predictions were aggregated to get the seasonal (monsoon or post monsoon) rainfall. The drought index simulations considered the datasets at monthly time scales, input selection is made by using PACF function and modeling is done by following MEMD-SLR and MEMD-GP method, along with their standalone counterparts. The analysis proved the superiority of MEMD-GP approach over the traditional linear and nonlinear models and marginal performance improvement over the MEMD-SLR approach. For the monthly inflow predictions, lagged inflow and lagged values of ENSO and EQUINOO were used. The MEMD-SLR method showed improved predictive capabilities of inflows into Hirakud reservoir when compared with MLR and data driven methods such as M5Model tree, GP and ANN while following a hindcasting scheme. The method was also found trustworthy in capturing the extreme flows (low and high) which are having significant impact in water resources planning. There found a clear evidence of mean and variance nonstationarity in the daily datasets of streamflow and SSF from Kallada river in Kerala in 2002–2012 period, in which a drastic change is noticed since 2009. The MEMD-SLR method exhibited superior performance over ANN, GP, MT and MLR methods in daily SSF prediction of Kallada river for validation data of 2009–2012 period, which clearly showed its potential in dealing with mean and variance nonstationarity.

REFERENCES

Adarsh, S. 2017. *Multiscale characterization of hydrologic time series using mathematical transforms*. Ph.D. thesis, IIT Bombay, India.

Adarsh, S., Janga Reddy, M. 2015. Trend analysis of rainfall in four meteorological subdivisions of southern India using non-parametric methods and wavelet analysis. *International Journal of Climatology* **35**(6): 1107–1124.

Adarsh, S., Janga Reddy, M. 2016. Analyzing the hydro-climatic teleconnections of summer monsoon rainfall in Kerala, India using Multivariate Empirical Mode Decomposition and Time Dependent Intrinsic Correlation. *IEEE Geoscience and Remote Sensing Letters* **13**(**9**): 1221–1225.

Adarsh, S., Janga Reddy, M. 2019. Multiscale characterization and prediction of reservoir inflows using MEMD-SLR coupled approach. *Journal of Hydrologic Engineering ASCE* https://doi.org/10.1061/(ASCE)HE.1943-5584.0001732 24(1).

Barnhart, B.L., Eichinger, W.E. 2011. Empirical mode decomposition applied to solar irradiance, global temperature, sunspot number and CO_2 concentration data. *Journal of Atmospheric and Solar-Terrestrial Physics* **73**(2011): 1771–1779.

Bazrafshan, J., Hejabi, S., Rahimi, J. (2014). Drought monitoring using the Multivariate Standardized Precipitation Index (MSPI). *Water Resources Management* **28**(4): 1045–1060.

Belayneh, A., Adamowski, J., Khalil, B., Ozga-Zielinskib, B. 2014. Long-term SPI drought forecasting in the Awash river basin in Ethiopia using wavelet neural network and wavelet support vector regression models. *Journal of Hydrology* **508**: 418–429.

Belayneh, A., Adamowski, J., Khalil, B. 2016a. Short-term SPI drought forecasting in the Awash River Basin in Ethiopia using wavelet transforms and machine learning methods. *Sustainable Water Resources Management* **2**: 87–101.

Belayneh, A., Adamowski, J., Khalil, B., Quilty, J. 2016b. Coupling machine learning methods with wavelet transforms and the bootstrap and boosting ensemble approaches for drought prediction. *Atmospheric Research* **172–173**: 37–47.

Bhattacharya, B., Solomatine, D.P. 2006. Machine learning in sedimentation modeling. *Neural Networks* **19**(2): 208–214.

Choubin, B., Malekian, A., Golsha, M. 2016. Application of several data-driven techniques to predict a standardized precipitation index. *Atmósfera* **29**(2): 121–128.

Gadgil, S., Vinayachandran, P.N., Francis, P.A., Gadgil, S. 2004. Extremes of the Indian summer monsoon rainfall, ENSO and equatorial Indian Ocean oscillation. *Geophysical Research Letters* 31 L12213 doi:10.1029/2004GL019733.

Ganguli, P., Reddy, M.J. 2013. Evaluation of trends and multivariate frequency analysis of droughts in three meteorological subdivisions of Western India. *International Journal of Climatology* **34**(3): 911–928.

Haji, M.S., Mirbagheri, S.A., Javid, A.H., Khezri, M., Najafpour, G.D. 2014. Suspended sediment modelling by SVM and wavelet. *Gradevinar* **66**(3): 211–223.

Hu, W., Si, B-C. 2013. Soil water prediction based on its scale-specific control using multivariate empirical mode decomposition. *Geoderma* **193–194**: 180–188.

Huang, N.E, Wu, Z., Long, S.R., Arnold, K.C., Blank, K., Liu, T.W. 2009b. On instantaneous frequency. Advances in Adaptive Data Analysis **1**(2): 177–229.

Huang, Y., Schmitt, F.G, Lu, Z., Liu, Y. 2009a. Analysis of daily river flow fluctuations using empirical mode decomposition and arbitrary order Hilbert spectral analysis. *Journal of Hydrology* **373**: 103–111.

Kashid, S.S., Maity, R. 2012. Prediction of monthly rainfall on homogeneous monsoon regions of India based on large scale circulation patterns using genetic programming. *Journal of Hydrology* **454**: 26–41.

Kim, T., Valdes, J.B. 2003. Nonlinear model for drought forecasting based on a conjunction of wavelet transforms and neural networks. *Journal of Hydrologic Engineering* **8**: 319–328.

Kisi, Ö., Yuksel, I., Dogan, E. 2008. Modelling daily suspended sediment of rivers in Turkey using several data-driven techniques. *Hydrological Sciences Journal* **53**(6): 1270–1285.

Krishnakumar, K.N., Rao, G.S.L.H.V.P., Gopakumar, C.S. 2009. Rainfall trends in twentieth century over Kerala, India. *Atmospheric Environment* **43**: 1940–1944.

Kumar, K.N., Rajeevan, M., Pai, D.S., Srivastava, A.K., Preethi, B. 2013. On the observed variability of monsoon droughts over India. *Weather and Climate Extremes* **1**: 42–50.

Lafdani, K.E, Nia, M.A., Ahmadi, A. 2013. Daily suspended sediment load prediction using artificial neural networks and support vector machines. *Journal of Hydrology* **478**: 50–62.

Maity, R., Nagesh Kumar, D. 2006a. Bayesian dynamic modeling for monthly Indian summer monsoon rainfall using El Niño-Southern Oscillation (ENSO) and Equatorial Indian Ocean Oscillation (EQUINOO). *Journal of Geophysical Research – Atmospheres* **111**, D07104, doi:10.1029/2005JD006539.

Maity, R., Nagesh Kumar, D. 2006b. Hydroclimatic association of monthly summer monsoon rainfall over India with large-scale atmospheric circulation from tropical Pacific Ocean and Indian Ocean region. *Atmospheric Science Letters* **7**(4): 101–107.

Maity, R., NageshKumar, D. 2008. Basin-scale streamflow forecasting using the information of large-scale atmospheric circulation phenomena. *Hydrological Processes* **22**(5): 643–650.

Maity, R., Nagesh Kumar, D. 2009. Hydroclimatic influence of large-scale circulation on the variability of reservoir inflow. *Hydrological Processes* **23**(6): 934–942.

Mallya, G., Mishra, V., Niyogi, D., Tripathi, S., Govindaraju, R.S. 2016. Trends and variability of droughts over the Indian monsoon region. *Weather and Climate Extremes* **12**: 43–68.

McKee, T.B., Doesken, D.J., Kleist J. 1993. The relationship of drought frequency and duration to time scales. *Proceedings of the 8th Conference on Applied Climatology*, vol. 17, American Meteorological Society, Boston, MA, USA, pp. 179–184.

Mishra, A.K., Singh V.P. 2011. Drought modeling – A review. *Journal of Hydrology* **403**: 157–175.

Nourani, V., Andalib, G. 2015. Daily and monthly suspended sediment load predictions using wavelet based artificial intelligence approaches. *Journal of Mountain Science* **2**(1): 85–100.

Pai, D.S., Sridhar, L., Guhathakurta, P., Hatwar, H.R. 2011. District wise drought climatology of the Southwest monsoon season over India based on Standardized Precipitation Index. *Natural Hazards* **59**: 1797–1813.

Rajaee, T. 2011. Wavelet and ANN combination model for prediction of daily suspended sediment load in rivers. *Science of the Total Environment* **409**: 2917–2928.

Rajaee, T., Mirbagheri, S.A., Zounemat-Kermani, M., Nourani, V. 2009. Daily suspended sediment concentration simulation using ANN and neuro-fuzzy models. *Science of the Total Environment* **407**: 4916–4927.

Rilling, G., Flandrin, P., Goncalves, P. 2003. On empirical mode decomposition and its algorithms, In *Proceedings of IEEE-EURASIP Workshop on Nonlinear Signal and Image Processing NSIP-03*, Grado (Italy), pp. 8–11.

Sahai, A.K, Soman, M.K., Satyan, V. 2000. All India summer monsoon rainfall prediction using an artificial neural network. *Climate Dynamics* **16:** 291–302.

Senthil Kumar, A.R., Ojha, C.S.P., Goyal, M., Singh, R., Swamee, P. 2012. Modeling of suspended sediment concentration at Kasol in India using ANN, fuzzy logic, and decision tree algorithms. *Journal of Hydrologic Engineering* **17**(3): 394–404.

Singh, K.K., Pal, M., Singh, V.P. 2010. Estimation of mean annual flood in Indian catchments using back-propagation neural network and M5 Model tree. *Water Resources Management* **24**: 2007–2019.

Singh, P., Borah, B. 2013. Indian summer monsoon rainfall prediction using artificial neural network. *Stochastic Environmental Research and Risk Assessment* **27**(7): 1585–1599.

Thomas, J., Prasannakumar, V. 2016. Temporal analysis of rainfall (1871–2012) and drought characteristics over a tropical monsoon-dominated State (Kerala) of India. *Journal of Hydrology* **534**: 266–280.

Thomas, T., Nayak, P.C., Ghosh, N.C. 2015. Spatio-temporal analysis of drought characteristics in the Bundelkhand region of Central India using the standardized precipitation index. *Journal of Hydrologic Engineering* http://dx.doi.org/10.1061/(ASCE) HE.1943–5584.0001189.

Zargar, A., Sadiq, R., Naser, B., Khan, F.I. 2011. A review of drought indices. *Environmental Reviews* **19**: 333–349.

Zounemat-Kermani, M., Kisi, O., Adamowski, J., Ramezani-Charmahineh, A. 2016. Evaluation of data driven models for river suspended sediment concentration modeling. *Journal of Hydrology* **535** (2016): 457–472.

7 Summary and Recommendations

Trend analysis of rainfall over different regions in India using DWT and EMD are proven to be robust in capturing the inherent nonlinear trend in different time series along with the well-debated significantly increasing trend in postmonsoon rainfall in Kerala subdivision. CEEMDAN-based decomposition is found to be efficient in terms of computational cost and quality of decomposition when compared with EEMD-based decomposition while applied for different applications. The MEMD-TDIC analysis for hydroclimatic teleconnections of ISMR and reservoir inflows is successful in detecting the nature (positive or negative) and strength of associations in the teleconnections, which displayed distinct differences with the time scale. MEMD-TDIC analysis showed that the hydroclimatic teleconnections are associated with localized alterations in the nature of associations over the time domain. This method is found to be successful in capturing the above normal rainfall of 1997–1998 below normal rainfall 2002–2003 period, the two well-debated events modulated by in the history of Indian climatology. The investigation of streamflow-sediment links using TDIC analysis showed that the link is fairly consistent (long range positive) at annual and intra-annual time scale irrespective of the location, while for longer time scales, the characteristics of such associations are location-specific. The climatic oscillations are dominant in hydrological variability of upper reaches of the basin, while human intervention plays a major role in deciding the variability at stations located at downstream reaches of the basin. The proposed MEMD-based hybrid approaches displayed substantial improvements in simulation of hydrological variables of diverse degree of complexity.

Most of the EMD-based hybrid decomposition models (or its variants) performed better than their standalone counterparts in hydrologic simulations, which may be because of the improved generalization capabilities resulted from the inclusion of information pertaining to different process scales during the calibration phase. However, more experiments on the scheme used to couple MEMD with different regression methods, division of calibration and validation datasets, the treatment of issues such as end effect through extension methods etc. are to be solicited to equip the method for real-time forecasting of hydrologic variables. The HHT-based methods have not applied much in the field of groundwater and water quality modeling, so there is a high potential to explore these methods via hybrid AI models involving EMD and other variants. Many of the past applications of EMD are mainly focused on temporal pre-processing, for extracting the underlying features and denoising time series. It could be useful to explore the use of the HHT for pre-processing of spatial data (e.g., digital elevation model), which may have a high impact on accuracy of hydrological modeling.

193

The MEMD-TDIC approach is proven to be one of the potential alternative to wavelet coherence to perform multiscale teleconnection studies, but the extensions works, such as applications using Time Dependent Intrinsic Cross Correlation (TDICC), should be investigated for hydrological applications is seriously missing in literature. The studies on denoising, synthetic data generation, complexity description of hydrologic series, AOHSA for fractal or multifractality description in hydrology, EMD/MEMD based multifractal or scaling analysis, etc. are very limited. The concurrent impacts of multiple climatic variables on hydrologic variability should be addressed using the techniques such as multiple wavelet coherence. Thus, there is ample scope for conducting more research works on different issues and to explore the complete potential of multiscale spectral analysis in hydrology.

Appendix A: Methods for Trend Analysis

Trends analysis of hydrologic time series is an important step in its characterization. Apart from the linear trend fitting, there are large number of parametric and non-parametric methods for analyzing trend of time-series data (Sang et al. 2013; Sonali and Nagesh Kumar 2013). Also, the nonlinear trend estimation is gaining popularity nowadays, which is capable to trace the slow/fast variation over smaller time domains (Wu et al. 2007; Unnikrishnan and Jothiprakash 2015). This section presents two of the most popular nonparametric methods of trend analysis–Sen's slope and Mann-Kendall (MK) methods along with the sequential version of the MK test used for estimation of sequential changes in trend.

A.1 SEN'S SLOPE METHOD

Sen's method (Sen 1968) is useful for estimating the slope of a linear trend and it has been widely used for determining the magnitude of trend in hydrometeorological time series (Lettenmaier et al. 1994; Yue and Hashino 2003; Partal and Kahya 2006; Jain and Kumar 2012; Jain et al. 2013). In this method, the slopes (m_i) of all data pairs are first calculated by:

$$m_i = \frac{(x_j - x_k)}{(j-k)} \text{ for } i = 1, 2, ..., N \qquad \text{(A.1)}$$

where N is the number of data points in the time series; x_j and x_k are data values at time j and k ($j>k$), respectively. The median of these N values of m_i is Sen's estimator of slope, which is calculated as follows:

$$\beta = \begin{cases} m_{\frac{N+1}{2}} & \textit{if } N \textit{ is odd} \\ \frac{1}{2}\left(m_{\frac{N}{2}} + m_{\frac{N+1}{2}} \right) & \textit{if } N \textit{ is even} \end{cases} \qquad \text{(A.2)}$$

A positive value of β indicates an upward (increasing) trend and a negative value indicate a downward (decreasing) trend.

A.2 MANN-KENDALL TEST

To ascertain the presence of statistically significant trend in hydroclimatic variables such as temperature, precipitation, and streamflow, etc., the nonparametric Mann-Kendall (MK) test (Mann 1945; Kendall 1975) has been employed by a number of researchers in the past (Yue et al. 2003; Burn et al. 2004; Pal and Al-Tabbaa 2009, 2011; Jain and Kumar 2012; Raj and Azeez 2012; Jain et al. 2013). The MK method helps to find the presence of a trend in a time series. The MK test has an advantage that it is applicable for data with outliers because its statistic is based on the sign of differences, not directly based on the values of the variable. Also, the MK test is applicable to the detection of a monotonic trend in a time series (Pal and Al-Tabbaa 2009).

The test is based on the statistic S, defined as follows:

$$S = \sum_{i=1}^{N-1} \sum_{j=i+1}^{N} \text{sgn}(x_j - x_i) \tag{A.3}$$

where N is the number of data points, x_j and x_i are data values at time j and i ($j > i$), respectively. This statistic represents the number of positive differences minus the number of negative differences for all the differences considered.

$$\text{Denoting } \delta = (x_j - x_i)$$

$$\text{sgn}(\delta) = \begin{cases} 1 & \text{if } \delta > 0 \\ 0 & \text{if } \delta = 0 \\ -1 & \text{if } \delta < 0 \end{cases} \tag{A.4}$$

For large samples ($N > 10$), the sampling distribution of S is assumed to be normally distributed with zero mean and variance as follows:

$$Var(S) = \frac{N(N-1)(2N+5) - \sum_{k=1}^{n} t_k(t_k - 1)(2t_k + 5)}{18} \tag{A.5}$$

where n is the number of tied (zero difference between compared values) groups and t_k the number of data points in the kth tied group.

The Z-statistic or standard normal deviate is then computed by using equation:

$$Z = \begin{cases} \dfrac{S-1}{\sqrt{Var(S)}} & \text{if } S > 0 \\ 0 & \text{if } S = 0 \\ \dfrac{S+1}{\sqrt{Var(S)}} & \text{if } S < 0 \end{cases} \tag{A.6}$$

A noticeable weakness of the MK test is that it does not account for serial correlation, which is very often found in precipitation and streamflow data (Hamed and

Rao 1998; Partal and Küçük 2006). The presence of an autocorrelation in a dataset may lead to inaccurate interpretations of the MK test. A time series exhibiting positive autocorrelation causes the effective sample size to be less than the actual sample size, thereby increasing the variance and the possibility of detecting significant trends when in fact, there are no trends (Hamed and Rao 1998; Ehsanzadeh et al. 2011). On the contrary, the existence of negative autocorrelation in a time series enhances the possibility of accepting the null hypothesis (absence of significant trends), when actually there are significant trends (Ehsanzadeh et al. 2011). Since using the original MK for autocorrelated data underestimates (or overestimates) the variance of the data, the calculation of the variance of the test statistic S is altered as given by an empirical formula proposed by Hamed and Rao (1998).

The variance of the test statistic can be computed as

$$Var(S) = \frac{n(n-1)(2n+5)}{18}\left(\frac{n}{n^*}\right) \tag{A.7}$$

where n is the actual number of observations and n^* is the effective number of sample size needed in order to account for the autocorrelation factor in the dataset. A correction associated with the autocorrelation of the data is applied in such cases. Empirically, the correction is expressed by Hamed and Rao (1998) as follows:

$$\left(\frac{n}{n^*}\right) = 1 + \frac{2}{n(n-1)(n-2)}\sum_{i=1}^{n-1}(n-i)(n-i-1)(n-i-2)\rho(i) \tag{A.8}$$

where

$\rho(i)$ is the autocorrelation function of the ranks of the observations.

Here, if the computed value of $|Z| > Z_{\alpha/2}$, then the null hypothesis of no trend is rejected at α level of significance in a two-sided test (i.e., the trend is significant). A positive value of Z indicates an increasing trend, and a negative value of Z indicates a decreasing trend.

A.3   SEQUENTIAL MANN-KENDALL (SQMK) TEST

The Mann-Kendall test helps to find monotonic trend in a time series. However, in hydroclimatic time-series data, the identification of starting time period of the significant trends is also of great interest. In addition, the determination of changes of trend over time is important in any trend detection study. The sequential Mann-Kendall (SQMK) test is particularly useful for such change detection analysis.

The SQMK test (Modarres and Sarhadi 2009; Sneyers 1990), which is progressive and retrograde analyzes of the Mann-Kendall test, will produce sequential values of standardized variables with zero mean and unit standard deviation. The sequential values are calculated for the forward series ($u(t)$), and backward series ($u'(t)$) in the basic framework of the Mann-Kendall test. The SQMK test allows for the detection of approximate beginning of a change. A detailed description of this method can be found elsewhere (Gerstengarbe and Werner 1999; Sneyers 1990).

The point at which the forward series crosses backward series is considered as potential trend turning point within the time series. Whenever the progressive ($u(t)$) or retrograde ($u'(t)$) row exceeds certain limits, before and after the crossing point, the null hypothesis of the sampled time series has no change points must be rejected, and this trend turning point may be significant at a particular significance level (i.e., say 5% significance level). The procedure of SQMK test involves the following steps:

1. The values of the original series (say) X_i are replaced by their ranks r_i, arranged in ascending order
2. The magnitudes of r_i, ($i = 1,2,...,n$) are compared with r_j, ($j = 1,2,...,i-1$); and at each comparison, the number of cases $r_i > r_j$ are counted and denoted by n_i
3. A statistic t_i is defined as follows:

$$t_i = \sum n_i$$

4. The mean and variance of the test statistic are computed as:

$$E(t_i) = \frac{i(i-1)}{4} \text{ and } Var(t_i) = \frac{i(i-1)(2i+5)}{72}$$

5. The sequential values of the statistic $u(t_i)$ can then be computed as:

$$u(t_i) = \frac{\left[t_i - E(t_i)\right]}{\sqrt{Var(t_i)}}$$

where $u(t_i)$ is a standardized variable that has zero mean and unit standard deviations. Therefore, its sequential behavior fluctuates around zero level. In the same way, $u'(t_j)$ is calculated starting from the end of the series (retrograde series).

If the intersection of $u(t)$ and $u'(t)$ occur within ±1.96 (that corresponds to the bounds at 5% significance level) of the standardized statistic, a detectable change has occurred at that point along the temporal time series (Olmo and Arboledas 1995). The change can be detected by Kendall coefficient (t_i) and when a time series shows a significant trend, the period from which the trend is noticeable can be obtained effectively by this test.

REFERENCES

Burn, D.H., Cunderlik, J.M., Pietroniro, A. (2004). Hydrological trends and variability in the Liard river basin. *Hydrological Sciences Journal* **49**: 53–67.
Ehsanzadeh, E., Ouarda, T.B.M.J., Saley, H.M. (2011). A simultaneous analysis of gradual and abrupt changes in Canadian low streamflows. *Hydrological Processes* **25**(5): 727–739.
Gerstengarbe, F.W., Werner, P.C. (1999). Estimation of the beginning and end of recurrent events within a climate regime. *Climate Research* **11**: 97–107.

Hamed, K.H., Rao, A.R. (1998). A modified Mann-Kendall trend test for autocorrelated data. *Journal of Hydrology* **204**: 182–196.

Jain, S.K., Kumar, V. (2012). Trend analysis of rainfall and temperature data for India-A Review. *Current Science* **102**(1): 37–49.

Jain, S.K., Kumar, V., Saharia, M. (2013). Analysis of rainfall and temperature trends in northeast India. *International Journal of Climatology* **33**(4): 968–978.

Kendall, M.G. (1975). *Rank Correlation Methods*. Charles Griffin: London, UK.

Lettenmaier, D.P, Wood, E.F, Wallis, J.R. (1994). Hydro-climatological trends in the continental United States (1948–88). *Journal of Climate* **7**: 586–607.

Mann, H.B. (1945). Non-parametric tests against trend. *Econometrica* **13**: 245–259.

Modarres, R., Sarhadi, A. (2009). Rainfall trends analysis of Iran in the last half of the twentieth century. *Journal of Geophysical Research* **114**: D03101, doi:10.1029/2008JD010707.

Olmo, F.J, Alados-Arboledas, L. (1995). Pinatubo eruption effects on solar radiation at Almeria (36.83°N, 2.41°W). *Tellus B* **47**: 602–660.

Raj, P.P.N., Azeez, P.A. 2012. Trend analysis of rainfall in Bharathapuzha River basin, Kerala, India. *International Journal of Climatology* **32**(4): 533–553.

Pal, I., Al-Tabbaa, A. (2009). Trends in seasonal precipitation extremes-An indicator of climate change in Kerala, India. *Journal of Hydrology* **367**: 62–69.

Pal, I., Al-Tabbaa, A. (2011). Monsoon rainfall extreme indices and tendencies in Kerala, India for 1954–2003. *Climate Change* **106**: 407–419.

Partal, T., Kahya, E. (2006). Trend analysis in Turkish precipitation data. *Hydrological Processes* **20**: 2011–2026.

Partal, T., Küçük, M. 2006. Long-term trend analysis using discrete wavelet components of annual precipitations measurements in Marmara region (Turkey). *Physics and Chemistry of the Earth* **31:** 1189–1200.

Sang, Y.F., Wang, Z., Liu, C. (2013). Discrete wavelet-based trend identification in hydrologic time series. *Hydrological Processes* **27**: 2021–2031.

Sen, P.K. 1968. Estimates of the regression co-efficient based on Kendall's tau. *Journal of the American Statistical Association* **63**: 1379–1389.

Sneyers, R. (1990). *On the statistical analysis of series of observations. WMO technical note 143*. WMO No. 415, TP-103, Geneva, World Meteorological Organization, p. 192.

Sonali, P., Nagesh Kumar, D. (2013). Review of trend detection methods and their application to detect temperature change in India. *Journal of Hydrology* **476**: 212–227.

Unnikrishnan, P., Jothiprakash, V. (2015). Extraction of non-linear trends using singular spectrum analysis. *Journal of Hydrologic Engineering*, 10.1061/(ASCE)HE.1943–5584.0001237, 05015007.

Wu, Z., Huang, N.E., Long, S.R., Peng, C.K. (2007). On the trend, detrending and variability of nonlinear and non-stationary time series. *Proceedings of National Academy of Science USA*, **104**: 14889–14894.

Yue, S., Hashino, M. (2003). Temperature trends in Japan: 1900–1990. *Theoretical and Applied Climatology* **75**: 15–27.

Yue, S ., Pilon, P., Phinney, B. (2003). Canadian stream flow trend detection: impacts of serial and cross-correlation. *Hydrological Sciences Journal* **48**: 51–63.

Appendix B: Scale Invariance Theory

Scaling implies that the statistical properties of the process observed at various scales are governed by the same relationship (Olsson 1998; Olsson and Berndtsson 1998). Let $X(t)$ and $X(\lambda t)$ denotes the observations (time series) at two distinct time scales t and λt, where λ is the scale conversion factor (a positive quantity). If $X(t)$ is scaling, then there exists a function $f(\lambda)$ such that

$$X(t) \underset{dist}{=} f(\lambda) \, X(\lambda t) \tag{B.1}$$

where $f(\lambda) = \lambda^{-\beta}$, $\underset{dist}{=}$ denote the equality of the probability distribution of the two random variables $X(t)$ and $X(\lambda t)$ and β is the scaling exponent, also known as the Mandelbrot–Kahane–Peyriere (MKP) Function (Mandelbrot 1974). In other words,

$$X(t) \underset{dist}{=} \lambda^{-\beta} X(\lambda t) \tag{B.2}$$

This is known as strict sense simple scaling (SSSS), which states that "equality in the probability distribution of the variable observed at two different time scales holds" (Gupta and Waymire 1990). This implies that the quantiles and raw moments of any order are also scale invariant, i.e., $E[X(t)^q] = \lambda^{-\beta q} E[X(\lambda t)^q]$ (referred to as wide sense simple scaling, WSSS), where q is the moment order. In simple scaling, the slope β can be obtained from the slope of the linear regression relation between log transformed values of $E[(X(\lambda t)^q]$ and the scale conversion factor λ for different orders of moment.

In the context of developing IDF curves of shorter duration rainfall from coarse resolution data, the usefulness of scaling theory can be described as follows. According to the scaling theory, a random variable, annual maximum rainfall intensity I_d obeys the simple scaling properties if it obeys the following (Gupta and Waymire 1990):

$$I_d = \left(\frac{D}{d}\right)^{-\beta} I_D \tag{B.3}$$

where D is the aggregated (coarse) time duration, which can be related to the duration d by defining the scale conversion factor $\lambda_d = \dfrac{D}{d}$

Hence, by scaling theory,

$$I_d = \lambda_d^{-\beta} I(\lambda_d d) \tag{B.4}$$

and

$$E[I(d)^q] = \lambda_d^{-\beta q} E[I(\lambda_d d)^q] \tag{B.5}$$

The assumption of WSSS enables us to derive the IDF relationships for subdaily durations from coarse resolution (daily or monthly) time-series data. Therefore, the q order probability weighted moments (PWMs) of rainfall intensity time series for various durations also follow scaling property as follows:

$$M_d^q = (\lambda_d)^{-\beta} M_D^q \tag{B.6}$$

where M_d^q and M_D^q are the PWM of intensity series of duration d and D, respectively. The PWM (Greenwood et al. 1979; Hosking 1986) is defined as

$$M^{i,j,s} = E[I_d^i F_{I_d}^j (1 - F_{I_d})^s] = \int_0^1 I_d^i F_{I_d}^j (1 - F_{I_d})^s dF_{I_d}$$

where F_{I_d} is the cumulative distribution function (CDF) of I_d. Adopting $i = 1, j = q$ and $s = 0$ (which are linear in I_d) is generally sufficient for parameter estimation. Thus, the left-hand side of the Equation (B.7) becomes $M^{1,q,0}$, which is generally denoted as M^q.

REFERENCES

Greenwood, J.A., Landwehr, J.M., Matalas, N.C., Wallis, J.R. (1979). Probability weighted moments: Definition and relation to parameters of several distributions expressible in inverse form. *Water Resources Research* **15**(5): 1049–1054.

Gupta, V.K., Waymire, E. (1990). Multiscaling properties of spatial rainfall and river flow distributions. *Journal of Geophysical Research* **95**: 1999–2009.

Hosking, J.R.M. (1986). *The theory of probability weighted moments*. Research Report RC 12210, IBM Research Division, Yorktown Heights, NY.

Mandelbrot, B.B. (1974). Intermittent turbulence in self-similar cascades: divergence of high moments and dimension of the carrier. *Journal of Fluid Mechanics* **62**: 331–358.

Olsson, J. (1998). Evaluation of a scaling cascade model for temporal rainfall disaggregation. *Hydrology and Earth System Science* **2**: 19–30.

Olsson, J., Berndtsson, R. (1998). Temporal rainfall disaggregation based on scaling properties. *Water Science and Technology* **37**(11): 73–79.

Appendix C: Extreme Value Formulation for Developing IDF Curves

The general expression of CDF of the extreme value (EV) distribution is given by

$$F(x) = e^{-e^{\left[\frac{-(x-\alpha)}{\beta}\right]}} \quad -\infty < x < \infty \text{ and } \beta > 0 \tag{C.1}$$

where x is the random variable; α and β are the parameters of the distribution.

The CDF of the random variable i_d following EV distribution is given by:

$$P(I_d \leq i_d) = F_d(i_d) = 1 - \frac{1}{T} = e^{-e^{\left[\frac{-(i_d - \mu_d)}{\sigma_d}\right]}} \tag{C.2}$$

where i_d is the rainfall intensity for duration d, T is the return period of i_d; μ_d and σ_d are the mean and standard deviation of the rainfall intensity series i_d.

The logarithmic transformation of Equation (C.2) gives:

$$-\ln\left(1 - \frac{1}{T}\right) = e^{\left(\frac{-(i_d - \mu_d)}{\sigma_d}\right)} \tag{C.3}$$

$$\ln\left(-\ln\left(1 - \frac{1}{T}\right)\right) = \frac{-(i_d - \mu_d)}{\sigma_d} \tag{C.4}$$

$$i_d = \mu_d + \sigma_d\left(-\ln\left(-\ln\left(1 - \frac{1}{T}\right)\right)\right) \tag{C.5}$$

For simple scaling, the statistical properties of EV distribution for two different time scales d and D are related as (Menabde et al. 1999):

$$\mu_d = \left(\frac{D}{d}\right)^{-\beta} \mu_D \tag{C.6}$$

$$\sigma_d = \left(\frac{D}{d}\right)^{-\beta} \sigma_D \qquad (C.7)$$

Substituting the expressions (C.6) and (C.7) in Equation (C.5),

$$i_d = \frac{\mu + \sigma\left(-\ln\left(-\ln\left(1 - \frac{1}{T}\right)\right)\right)}{d^{-\beta}} \qquad (C.8)$$

where μ and α are the location parameter and scale parameter, respectively, given by $\mu = \mu_D (D)^{-\beta}$ and $\sigma = \sigma_D (D)^{-\beta}$

REFERENCE

Menabde, M., Seed, A., Pegram, C.G.S. (1999). A simple scaling model for extreme rainfall. *Water Resources Research* **35**(1): 335–339.

Index

Page numbers in italics indicate figures and in bold indicate tables on the corresponding pages.

For Product Safety Concerns and Information please contact our EU
representative GPSR@taylorandfrancis.com
Taylor & Francis Verlag GmbH, Kaufingerstraße 24, 80331 München, Germany